Measuring Intelligence, Surveillance, and Reconnaissance Effectiveness at the United States Central Command

DAVID LUCKEY, BRADLEY KNOPP, SASHA ROMANOSKY,
AMANDA WICKER, DAVID STEBBINS, CORTNEY WEINBAUM,
SUNNY D. BHATT, HILARY REININGER, YOUSUF ABDELFATAH,
SARAH HEINTZ

Prepared for the U.S. Central Command Directorate of Intelligence
Approved for public release; distribution unlimited

RAND NATIONAL DEFENSE RESEARCH INSTITUTE

For more information on this publication, visit www.rand.org/t/RR4360

Library of Congress Cataloging-in-Publication Data is available for this publication.
ISBN: 978-1-9774-0477-0

Published by the RAND Corporation, Santa Monica, Calif.

Cover images: AWACS Surveillance Aircraft: JamesReillyWilson/Getty Images/iStockphoto; Computer unit: RichardBaker/Alamy Stock Photo

Preface

This report is intended to assist the U.S. Central Command (CENTCOM) Directorate of Intelligence (J2) in improving the quality, effectiveness, and efficiency of its intelligence, surveillance, and reconnaissance (ISR) assessments. The purpose of this report is to help CENTCOM J2 develop repeatable, scalable, data-informed support for measuring the effectiveness of ISR with specific targets to evaluate current performance and plan for, influence, and resource future methodology and training. In this report, RAND Corporation researchers provide research and analysis to support CENTCOM J2 by assessing the effectiveness of its ISR resources. While this report has direct applicability to CENTCOM, it also has applications for the Joint Staff and other U.S. Department of Defense geographic and functional combatant commands. The observations, findings, and recommendations should also be of interest to organizations and stakeholders inside and outside the federal government who conduct intelligence activities of any type or who assess mission performance or effectiveness. The research reported here was completed in February 2020 and underwent security review with the sponsor and the Defense Office of Prepublication and Security Review before public release.

This research was sponsored by CENTCOM J2 and conducted within the Cyber and Intelligence Policy Center (CIPC) of the RAND National Defense Research Institute, a federally funded research and development center sponsored by the Office of the Secretary of Defense, the Joint Staff, the Unified Combatant Commands, the Navy, the Marine Corps, the defense agencies, and the defense intelligence enterprise.

For more information on the RAND Cyber and Intelligence Policy Center, see www.rand.org/nsrd/intel or contact the director (contact information is provided on the webpage).

Contents

Figures

Tables

Summary

The U.S. Central Command's (CENTCOM or "Command") Intelligence, Surveillance, and Reconnaissance (ISR) Assessments Directorate reviews and evaluates ISR operations (ops) by measuring ISR tasking and collection activities. The purpose of these tasking and collection assessments is to provide time-sensitive, integrated, and synchronized intelligence (INTEL) support to operations and policy staffs internally at the Command and externally to other defense and national stakeholders. CENTCOM uses this assessment capability to (1) enable a data-driven, fact-based discussion of ISR support; (2) drive Command and Joint Staff investment and disinvestment decisions; (3) inform resource allocation and process redesign, allocation, and apportionment of ISR assets; and (4) provide detailed evaluation of consumer satisfaction. To improve the quality of their analysis and better understand the output and outcome of ISR operations, the Assessments Directorate staff requested that RAND develop a methodology to measure the performance and effectiveness of ISR operations and activities. CENTCOM sought a methodology that would provide a repeatable, reproducible performance management process and framework that captures and monitors how CENTCOM ISR performs throughout its lifecycle of planning, tasking, execution, and reporting.

We developed measures of effectiveness (MoEs) and measures of performance (MoPs) by combining top-down and bottom-up approaches. The top-down approach started with the assumption that the mission support roles, sub-roles, and activities were defined in joint or service doctrine. The bottom-up approach assumed that the data already being collected by ISR platforms and sensors measure performance. We created a framework to apply the MoPs and MoEs to CENTCOM's ISR support mission. We reviewed CENTCOM's current databases and data capture and aggregation processes to ensure they are synchronized with the metrics we developed, and to ensure consistency in the collection of data that support long-term analysis.

We created a visualization tool to analyze the data and display the assessment results, which allow analysts and other stakeholders to make data-driven decisions within their realm of authority. Such a tool allows users to alter the parameters of the analysis and to see correlations that would not otherwise be easily visible and enhance decisionmaking.

Additionally, we developed repeatable, scalable, data-informed metrics to assist CENTCOM with improving the quality, effectiveness, and efficiency of ISR assessments. The data visualization tool expands the impact of these metrics by improving situational awareness and supporting decisionmaking. While we specifically analyzed the outputs and outcomes of CENTCOM ISR support operations, the methodology, findings, and visualization tool could also be applied to organizations and stakeholders inside and outside the federal government that conduct intelligence activities of any type or any organization that assesses mission performance or effectiveness.

We identified several obstacles to effective ISR assessment. There is a lack of alignment of program requirements with assessments. No single definition for airborne ISR capabilities or requirements exists in U.S. Department of Defense (DoD) doctrinal publications nor throughout the non-DoD literature we examined. Rather, many ISR definitions are meant to assist global force management in matching ISR platforms to specific combatant command (CCMD) missions. There is a lack of requirement-to-strategy linkage such that ISR platforms are rarely evaluated on the ability to produce intelligence that closes intelligence gaps; therefore, it is not possible to quantify the contribution ISR makes to CCMD decisionmaking or to the increased collective understanding of the operational environment. Higher-level assessment strategies require inputs from multiple stakeholders—e.g., intelligence analysts, planners, operators—at subordinate levels to ensure that a logical linkage between MoPs and MoEs exists. More granular system-level assessments might require the support of only three to five personnel, because those platforms are tied to more specific tasks. External stakeholders may have additional information that can provide additional inputs, adding increased value to MoE statements—or may complicate the assessment process because they view ISR through the lens of their own missions and have different interests in and concerns with the data analysis.

Another topic of concern is data latency, or the time delay between data collection, processing, and delivery to the end user. Data latency is critical when viewed through the lens of ISR; the loss of time or fidelity in data transfer could cost lives. Minimizing data latency through the integration of ISR requirements for ISR systems is important in cases where commanders require near real-time information to make critical decisions. The data latency effect can inhibit typical decisionmaking. Assessments that aim to inform either equipment acquisition or resource planning will require a different set of forward-looking metrics, perhaps reliant on alternative streams of data, than the metrics used to assess current operations.

We suggest that developing fewer simple metrics that are measurable, relevant, and clearly communicable across stakeholders would be more useful than developing a great number of platform-specific ones; however, defining metrics in this way can be challenging. Current disparate ISR metrics that lend themselves to consolidation are more likely to be measurable, relevant, and understood throughout ISR plans and operations. Consolidating a vast array of metrics or data elements into more-discrete, readily measurable data elements can inform a variety of assessments across organizations and operations. The unique nature of disparate measures, however, could be lost after consolidating discrete metrics or data elements. It is difficult to plan or evaluate results with metrics containing multiple discrete aspects. Moreover, the use of former data elements to inform the newly consolidated metric will likely morph during this process. Despite such difficulties, we suggest that the consolidation of similar metrics will enable more-effective ISR assessment processes.

There are three general takeaways from our review. First, higher-level assessment strategies usually require inputs from multiple stakeholders, while system-level assessments might require the support of only a few personnel. Second, assessments that aim to inform either equipment acquisition or resource planning will require a different set of forward-looking metrics than the metrics used to assess current operations. Third, similar data elements may inform a variety of assessments across organizations and operations, but how those data elements are used will vary. Here, the consolidation of similar metrics will assist ISR assessment processes moving forward.

When aligning program requirements to program assessments, we found that the most pertinent insight was that MoPs must focus on outputs, while MoEs need to measure mission outcomes. Both sets of metrics must also clearly be defined in the context of their use and should be interrogated to ensure that they can be gathered as needed, be accurate, and ensure that the metric design focuses on data and systems currently in use.

Generalizing and consolidating metrics will improve the overall assessment process. The proliferation of unique ISR platforms has increased in the post-2001 era, creating the perception that each platform must therefore require its own unique set of metrics to gauge success. As simple metrics have morphed to support more complicated or unique ISR platforms, however, it has become more difficult to compare effectiveness across platforms. Sets of MoEs and MoPs should be logically simplified enough to support general assessments across various platforms to ease the burden on those collecting and making sense of the data, and to improve the timeliness and quality of information that commanders require to make decisions.

Acknowledgments

The RAND National Defense Research Institute project team would like to thank Col Christopher Huisman, Christopher Renner, and Gary Graham from U.S. Central Command Directorate of Intelligence (CENTCOM J2) for their support for and assistance with this project. We would also like to thank our RAND colleagues Elizabeth Hammes, for assisting with our research; Katheryn Giglio, for communications expertise; Mike Decker and Bruce McClintock, for their quality assurance reviews; Rich Girven, for Cyber and Intelligence Policy Center leadership; and Jack Riley for NDRI leadership.

While we are thankful for the assistance provided by those listed and others, we ask that you attribute any errors or omissions solely to the authors.

Abbreviations

AOR	area of responsibility
BDA	battle damage assessment
CCIR	commander's critical information requirement
CCMD	combatant command
CENTCOM	U.S. Central Command
DoD	U.S. Department of Defense
EEI	essential element of information
F3EAD	find, fix, finish, exploit, analyze, and disseminate
FDA	functional damage assessment
FPS	fixed point security
GAO	U.S. Government Accountability Office
HRI	human-robot interaction
HVI	high-value individual
IED	improvised explosive device
INTEL	intelligence
ISR	intelligence, surveillance, and reconnaissance
ITA	initial target assessment
J2	Directorate of Intelligence
JIPOE	joint intelligence preparation of the operational environment

JTCB	Joint Targeting Coordination Board
KNK	kinetic/nonkinetic
MoE	measure of effectiveness
MoP	measure of performance
NDRI	National Defense Research Institute
NOAA	National Oceanic and Atmospheric Administration
ODNI	Office of the Director of National Intelligence
OODA	observe, orient, decide, act
ops	operations
OSD	Office of the Secretary of Defense
OUSDI	Office of the Under Secretary of Defense for Intelligence
PDA	physical damage assessment
PED	processing, exploitation, and dissemination
PID	positive identification
PIR	priority intelligence requirement
PMESII	political, military, economic, social, information, and infrastructure
POA	pattern of activity
POL	pattern of life
STA	supplemental target assessment
TCPED	tasking, collection, processing, exploitation, and dissemination
TDN	target development nomination
TSA	target system assessment
USAF	U.S. Air Force

CHAPTER ONE

Introduction

Background

The U.S. Central Command (CENTCOM) directs and enables military operations (ops) and activities with allies and partners to increase regional security and stability in support of enduring U.S. interests. Its mission is complex and complicated. Since the aftermath of the terrorist attacks of September 11, 2001, CENTCOM has been charged with the responsibility of commanding multiple, often simultaneous, combat missions. During that time, confronting terrorism and defeating violent extremist groups was the primary objective of U.S. national military power. Seventeen years later, CENTCOM is still conducting multiple, active combat operations.[1]

The size of the Command's area of responsibility (AOR), its continuing operations in Afghanistan, Syria, and Iraq, and the threat of other regional conflicts demand continuous intelligence (INTEL) collection to ensure situational awareness. CENTCOM's intelligence, surveillance, and reconnaissance (ISR) activities have centered around Iraq and Afghanistan and have expanded to other regional issues.[2] ISR was used to identify Al-Qaida members and affiliates, then to support U.S. combat operations, then to support counterinsurgency operations, then to support stability operations, and now to monitor security around U.S. and coalition bases, identify threats to U.S. and coalition forces, and monitor threats from other regimes.

At present, the Command's AOR includes adversarial relationships among neighboring states, widespread ethnic and sectarian struggles, malign influence and desta-

---

1 Joseph Votel, "National Security Challenges and U.S. Military Activities in the Greater Middle East and Africa," testimony before the U.S. House of Representatives Armed Services Committee, Washington, D.C., March 7, 2013.

2 As defined by the U.S. Department of Defense (DoD) in Joint Publication 2-01 and the DoD *Dictionary of Military and Associated Terms* (as of July 2019), *intelligence, surveillance, and reconnaissance* is "1. An integrated operations and intelligence activity that synchronizes and integrates the planning and operation of sensors, assets, and processing, exploitation, and dissemination systems in direct support of current and future operations. 2. The organizations or assets conducting such activities. Also called ISR. See also intelligence; intelligence, surveillance, and reconnaissance visualization; reconnaissance; surveillance" (U.S. Joint Chiefs of Staff, *Joint and National Intelligence Support to Military Operations*, Joint Publication 2-01, July 5, 2017c).

bilizing activities, cyber-based threats, and growing arsenals of sophisticated conventional weapons and weapons of mass destruction that all combine to imperil enduring U.S. vital national interests, as well as those of trusted U.S. partners and allies.[3] Command-owned ISR assets play a major role in this effort.

CENTCOM's Office of ISR Assessments reviews and evaluates the full spectrum of ISR operations by measuring ISR tasking and collection activities. CENTCOM ISR tasking and collection assessments aim to improve time-sensitive, integrated, and synchronized intelligence support to operations and policy staffs at the Command and to national customers. CENTCOM ISR staff provided RAND Corporation researchers with five specified roles that ISR is tasked to support. These roles are defined in Table 1.1.

Purpose

The CENTCOM J2 ISR Assessment Staff asked the RAND team to develop a methodology to measure the effectiveness of ISR operations and activities that would improve the quality of their analysis and their ability to understand the output and outcome of ISR operations. This assessment capability seeks to enable CENTCOM to have a data-driven, fact-based discussion of ISR support; drive investment and disinvestment decisions at the Command and Joint Staff level; and inform resource allocation and process

Table 1.1
CENTCOM ISR Roles and Definitions

Role	Definition
1. ISR support to fixed point security (FPS)	Support to surveillance in the immediate vicinity of a fixed location with a permanent or semipermanent presence of coalition forces, e.g., garrisons, forward operating bases, aerial or sea ports of debarkation. Preventative measures taken to mitigate hostile actions against United States, coalition, resources, facilities, and critical infrastructure.
2. ISR support to ops	Support to overwatch and force protection of maneuvering friendly forces (land, sea, or air). Includes dynamic targeting support to units conducting advise and assist missions.
3. ISR support to high-value individual (HVI)	Support to the development and refinement of HVI target nominations. Components: find, fix, track, target, engage, and assess.
4. ISR support to kinetic/nonkinetic (KNK) targeting	Support for the development and refinement of target nominations for Joint Targeting Coordination Board (JTCB) or similar body adjudication, operational strike planning, and battle damage assessment (BDA).
5. ISR support to intelligence (INTEL)	Support to intelligence production, joint intelligence preparation of the operational environment (JIPOE), and warning.

SOURCE: CENTCOM Directorate of Intelligence (J2) ISR Assessments Office and RAND analysis.

[3] CENTCOM, "Area of Responsibility," webpage, undated.

redesign, allocation and apportionment of ISR assets; and provide detailed evaluation of consumer satisfaction.

CENTCOM sought a methodology to provide a repeatable, reproducible performance management process and framework that capture and monitor how CENTCOM ISR performs throughout its lifecycle of planning, tasking, execution, and reporting. From previous work by the authors and others (see References), we assess that the critical element of any performance management framework is measuring output and outcomes. The ISR Assessment Staff focuses its output on the needs of its core customers: warfighters, the Joint Staff, the Office of the Under Secretary of Defense for Intelligence, the Office of the Secretary of Defense (OSD), and the Office of the Director of National Intelligence. For warfighters and policymakers, the most important output is high-quality data that can be folded into all-source intelligence analysis aimed at informing decisions on strategy and tactics. Other decisionmakers concerned with resource allocation, or allocation and apportionment of ISR capabilities, are more interested in data and analysis of the effectiveness of the actual ISR operations itself. Operators who staff and maintain ISR equipment focus on details of the operation of the ISR equipment so that it can be properly used and maintained. Measuring outputs and outcomes for this diverse group presents a challenging assessment problem.

This report describes the methodology we developed that will allow CENTCOM to measure the performance and effectiveness of ISR operations and activities. For the five major roles assigned to ISR support, we developed measures of effectiveness (MoEs) and measures of performance (MoPs), which aim to provide as objective and quantitative an assessment as possible.[4] Finally, we developed a data visualization tool that facilitates additional data correlations and provides analysts and managers with an easy-to-digest view of the information.

Assumptions and Limitations

The scope and methodology of this research are underpinned by several key assumptions.

First, this report focuses specifically on the assessment of ISR tasking and collection but excludes any consideration of processing, exploitation, and dissemination (PED). The ISR Assessments Office is responsible for assessing ISR collection in response to taskings and has the ability to collect and aggregate data appropriate to

[4] Equating MoEs with "outcomes" and MoPs with "outputs" is essential to understand the unique nature of metrics required for ISR assessments. Other publications, such as the *Commander's Handbook for Assessment Planning and Execution*, published by the Joint Staff J-7, offer different definitions that are more complex and do not resonate as well in the nuanced and specific context of ISR assessments at CENTCOM (*Commander's Handbook for Assessment Planning and Execution*, version 1.0, Suffolk, Va.: Joint Warfighting Center, Joint Staff J-7, Joint and Coalition Warfighting, September 9, 2011a).

assessment activities. The office has little capability to task for or collect PED data. At CENTCOM, PED remains an area that has not been fully explored by the ISR Assessments division. ISR Assessments could task subordinate units for data via the ISR Execute Order; however, they would first need to determine what data are required, for what purpose, and whether those data were already available via other sources.

Second, the methodology we created addresses CENTCOM-specific requirements. We suggest that this methodology could be adjusted to fulfill the specific requirements of other combatant command (CCMD) ISR operators if those organizations use the approach outlined in this report in the development of MoPs and MoEs. Additionally, this methodology could be scoped to assess more tactical ISR operations or to review more strategic ISR operations. This scoping, however, could potentially necessitate the development of additional, unique MoPs and MoEs.

Third, we made a fundamental assumption that ISR assets were being tasked to collect only information for which the systems were optimized from the outset of the project. This assumption underpins two important elements of the methodology: the collection requirements are data points and the requirements must be collectible by ISR sensors and platforms. One of the key components of data collected for ISR missions is the collection requirement itself. Aggregation and analysis of these collection requirements is critical to ISR performance assessments because they reflect the specific needs of the consumer. The requirement validation process would also serve to screen out any tasks that were beyond the collection capability of ISR sensors and platforms. The assignment to ISR of collection missions beyond its collection capabilities would dilute the quality of the data used to assess its performance and effectiveness. Additional analysis for this optimization could include a way to account for suboptimal collection approaches. It appears to us that it might make more sense to give credit for systems that provide some level of ISR response to collection requirements rather than excluding data that end up favoring the systems rather than the mission requirements. The analysis required to accomplish this assessment, however, is beyond the scope of our research. Our fourth assumption, detailed next, reinforces the need to account for unserved collection requirements. A mission need is indicated by a customer submitting a collection requirement. The ability to meet that customer's need should be assessed because, if only requirements that are within the capability of sensors and platforms are assessed, then the true needs of customers are not fully addressed.

Fourth, we assume here that mission collection requirements serve as a proxy for input and evaluation from consumers of the intelligence. Obtaining high-quality feedback from users of ISR data—analysts, policymakers and decisionmakers, operational commanders—is a difficult task at best. Our research showed that feedback data like these were infrequently obtained and not of great value in assessing the effectiveness of ISR collection; many consumers were unfamiliar with the sources of specific pieces of information that generally arrives as integrated, all-source reporting, and thus few consumers could comment definitively on the value of the ISR input. The collection requirement, however, is a validated statement of consumer needs. ISR collection requirements may not represent consumers' entire information needs, but rather only the portion of the need that is most available to ISR. The collection requirements, therefore, would reflect actual consumer needs and serve as the best possible proxy for use in assessments of ISR performance and effectiveness.

Report Structure

The RAND National Defense Research Institute used a mixed methodology to address the issues raised by CENTCOM. Following this chapter, Chapter Two introduces the literature we reviewed. We summarize previous RAND analysis and external analysis on performance management and ISR systems. Chapter Three discusses CENTCOM's ISR roles, sub-roles, and activities. Chapter Four presents our methodology for developing an assessment approach for CENTCOM. We assessed CENTCOM's five ISR roles, examined outputs and outcomes, and then identified MoPs and MoEs. Chapter Five summarizes our visualization of metrics. Chapter Six contains observations, findings, and conclusions.

This report also contains several appendixes. Appendix A contains a detailed literature review. Appendix B contains a blank version of the RAND-developed assessment framework. Appendix C houses the full assessment of the ISR mission, using the framework described in Appendix B. Appendix D provides views of the Tableau-based data visualization tool. Finally, we provide a list of all references used in this report.

CHAPTER TWO

Literature Review Results

CENTCOM's Office of ISR Assessments asked RAND to develop a methodology to evaluate CENTCOM's ISR activities by measuring tasking and collection operations within its AOR. We conducted a literature review to examine how similar organizations in government, academia, and the private sector carry out assessments of ISR platforms and sensors. This chapter provides a selection of our key findings on ISR-metric practices used by the wider ISR community. The entire literature review and our methodology are in Appendix A.

Themes

We found several commonalities to facilitate comparison across documents. First, many of the documents focused on operational definitions of ISR. Second, the literature provided general guidance on overall ISR assessment processes, though not with enough granularity to differentiate among various services. This could partly explain why CENTCOM has had difficulty creating meaningful MoEs and MoPs, despite the existence of a body of literature on this topic. Third, we found several obstacles to effective implementation of ISR assessments, including the alignment of program requirements to assessment process, issues with data latency, and the proliferation of redundant or disparate measures across platforms.

Brief Discussion of MoPs and MoEs in Use for ISR Assessment

Several government publications offer ISR assessment guidance for J2 mission planners, though not at level of granularity required for an effective ISR assessment framework.[1] Joint Publication 2-01 and the 2011 *Commander's Handbook for Persistent Sur-*

---

[1] For example, we reviewed other Joint Publications (Joint Fire Support, JIPOE, and Joint Planning) that do not offer effective assessment guidance. For example, see U.S. Joint Chiefs of Staff, *Joint Fire Support*, Joint Publication 3-09, December 12, 2014b. However, other publications, such as U.S. Joint

veillance offer the best baseline assessment guidance for J2 planning purposes.[2] Joint Publication 2-01 explains that the purpose of an ISR assessment is to "evaluate the performance of intelligence collection operations in order to improve collection effectiveness in meeting intelligence and operational requirements."[3] Within this context, such assessments can drive continuous improvement during the tasking, collection, processing, exploitation, and dissemination (TCPED) process. There are four suggested general ISR assessment categories for use by CCMDs: (1) MoPs, which should involve the evaluation of incremental activities geared toward task accomplishment; (2) MoEs, which help determine how well the overall mission was accomplished; (3) *formative assessments*, which serve as mission best practices to inform future sorties; and (4) *summative assessments*, which can aid in the commander's decision by documenting the overall contribution of ISR activities in meeting command priorities.[4] The literature review suggests that the use of these four categories will increase overall ISR assessment planning processes.

Findings

Our review helped to inform ISR roles and activities found in Chapter Three by explaining how metrics are used throughout DoD and the public and private sectors. This helped our team develop assessment mechanisms that would be most useful to the CENTCOM J2 staff. Subsequent findings related to lack of guidance, data latency, and disparate reporting and data streams led to the promulgation of our data visualization tool.

Our analysis revealed differences in ISR assessment at the data analysis level, along with considerable commonality among the concepts identified in DoD ISR doctrine. The first commonality is that higher-level assessment strategies typically require input from multiple internal and external stakeholders, while J2 system-level assessments draw on the support of three to five personnel. Second, assessments that aim to inform equipment acquisition or resource planning require a more complex method of understanding future requirements than current operational-level metrics allow.

A third commonality shows that while similar data elements can inform a variety of assessments across organizations and operations, how those data points are opera-

---

Chiefs of Staff, *Joint Intelligence Preparation of the Operational Environment*, Joint Publication 2-01.3, May 21, 2014a; U.S. Joint Chiefs of Staff, *Joint Planning*, Joint Publication 5-0, June 16, 2017a; and U.S. Joint Chiefs of Staff, *Geospatial Intelligence in Joint Operations*, Joint Publication 2-03, July 5, 2017b, helped to define our proposed assessment model in Chapter Four.

2 U.S. Joint Chiefs of Staff, 2017c; *Commander's Handbook for Persistent Surveillance*, version 1.0, Suffolk, Va.: Joint Warfighting Center, Joint Doctrine Support Division, June 20, 2011b, p. I-1.

3 U.S. Joint Chiefs of Staff, 2017c.

4 U.S. Joint Chiefs of Staff, 2017c.

tionalized varies, and that generalizing and consolidating metrics yielded optimum results among various assessment processes. The proliferation of ISR platforms has increased in the post-2001 era, and simple metrics have morphed to support more complicated ISR platforms or multiple operational objectives. For example, ISR operators may be focused on platform flight hours to enable planning for maintenance and equipment upgrade, while decisionmakers may be more concerned with the available flight hours that ensure the kind of persistent surveillance capability required to support their efforts. Here, the consolidation of like metrics will assist ISR assessment processes moving forward.

Finally, the literature suggests that MoPs must focus on outputs (usually a quantitative measure or binary), while MoEs need to measure mission outcomes (usually qualitative). MoPs and MoEs should be clearly defined within the context of their use and must be routinely tested to ensure that they can be gathered as needed (a reliability measure), remain accurate (a validity measure), and ensure that the metric focuses on data and systems currently in use (a feasibility measure).

CHAPTER THREE

CENTCOM ISR Roles, Sub-roles, and Activities

Overview

CENTCOM assigns five roles to ISR assets: (1) support to FPS; (2) support to HVI ops; (3) support to ops; (4) support to KNK targeting; and (5) support to INTEL. These roles were provided to us by CENTCOM and are shown in Figure 3.1.

For purposes of our analysis, we binned the five roles into three categories based on the different types of ISR support each requires:

1. ISR support to current ops
2. ISR support to future ops
3. ISR support to strategic planning.

This chapter explains how we used these roles to develop metrics. We started with joint and service-specific governance documents (e.g., regulations, instructions, and manuals), along with discussions with CENTCOM staff to deconstruct each specific *role* into *sub-roles*,[1] and (if necessary) these *sub-roles* were further broken down into *activities*. This hierarchy provided the framework required for detailed analysis. The terms *role*, *sub-role*, and *activity* are used throughout the remainder of this report to refer to the three levels of CENTCOM ISR tasks.

Figure 3.1
CENTCOM ISR Roles

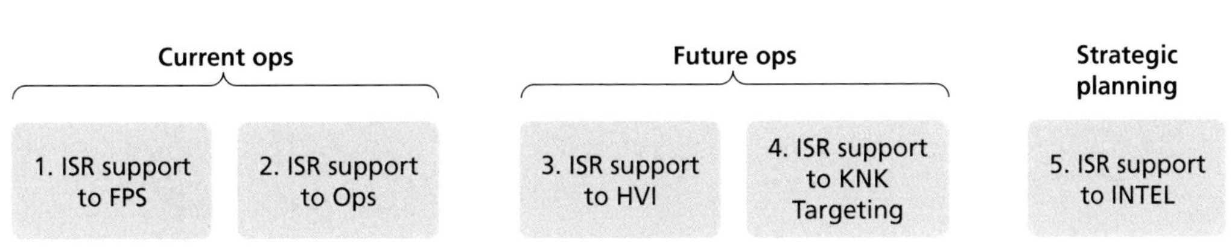

SOURCE: CENTCOM J2 ISR Assessments Office and RAND analysis.

---

[1] Doctrine includes intelligence tasks for each sub-role and activity, but the tasks were not tailored to unique ISR capabilities. The sub-roles and activities identified here are the most appropriate for ISR collection.

Current Operations

The first category includes ISR support to current operational activities. These current ops range from major regional deployments, such as Afghanistan and Iraq, to ops more limited in scope, such as noncombatant evacuation operations; peace operations; humanitarian assistance; personnel or equipment recovery operations; and chemical, biological, radiological, and nuclear response actions. ISR support to these ops requires agility in responding to near-term tasks that evolve as the mission proceeds.

We developed sub-roles and activities for FPS and ops, which are shown in Figure 3.2.

The first role, ISR support to FPS, is a security operation conducted to protect friendly forces, installations, routes, and actions within a specific area. Within CENTCOM, the FPS task is defined as a base or base camp defense operation. ISR's task is limited to a seven-day period and provides coverage of the facility.[2]

The second role, ISR support to ops, refers to intelligence collection and analysis capabilities that provide dedicated intelligence support to units in maneuver. For CENTCOM, this task comprises two specific sub-roles: provide intelligence information to units prior to and during ops (*overwatch*), and provide dynamic targeting of

Figure 3.2
ISR Support to Current Operations

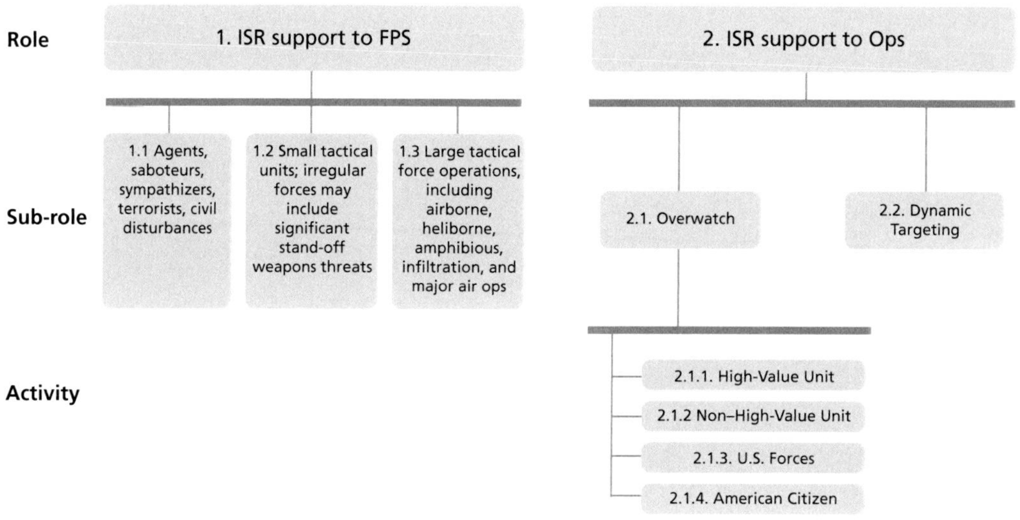

SOURCE: RAND analysis based on CENTCOM input.

---

2 Army Doctrine Publication 3-37, *Protection*, July 2019, supersedes ADRP 3-37, August 2012.

unplanned and unanticipated threats to operational units as necessary during ops.[3] Within overwatch, CENTCOM places different ISR priority on overwatch of different units to manage which ISR assets are tasked where. CENTCOM's prioritization is based on whether the operation involves a high-value unit, a non–high-value unit, other U.S. forces, or other American citizens.

Future Operations

The second category includes roles that are focused on support to future ops, including ISR support to HVI targeting and KNK targeting. These ops generally occur over a more extended period of time—weeks or months—than activities in current ops, which persist only as long as the operation is underway. These roles involve ISR in response to tasks that help intelligence support target development and selection processes. Once the target is selected and an operation is planned, ISR support is more characteristic of support to military ops, as noted above. The sub-roles and activities we developed for future ops are shown in Figure 3.3.

ISR support to high-value targets involves a person of interest who is identified, surveilled, tracked, influenced, or engaged. ISR plays a critical role in providing the persistent reconnaissance required to accurately find, fix, and track a target and to assess the impact of actions taken against that target.[4] We do not assess the *target* and *engage* components of the find, fix, track, target, engage, and assess cycle, because this is specific to weapons delivery and outside the scope of assessing ISR.

ISR support to KNK targeting refers to the development of target packages and the BDA conducted after an attack. ISR operations gather and process needed data and help improve the accuracy and extent of assessments that underpin target selection and validation.[5]

The first sub-role is support to deliberate target development, the process by which planned targets are prosecuted. This activity consists of four phased activities: Target Development Nomination (TDN), pattern of activity (POA) assessment, pattern of life (POL) assessment, and positive identification (PID) of the target.

The second sub-role is support to BDA. BDA compares postexecution results with the projected results generated during the target development. Comprehensive BDA requires a coordinated effort between intelligence and ops staff. *BDA* is defined in three phased activities:

---

3 Army Technical Publication 3-60, *Targeting*, May 2015.

4 Army Technical Publication 3-60, 2015.

5 U.S. Joint Chiefs of Staff, *Joint Targeting*, Joint Publication 3-60, January 31, 2013, Appendix D, The Targeting Assessment Process.

Figure 3.3
ISR Support to Future Operations

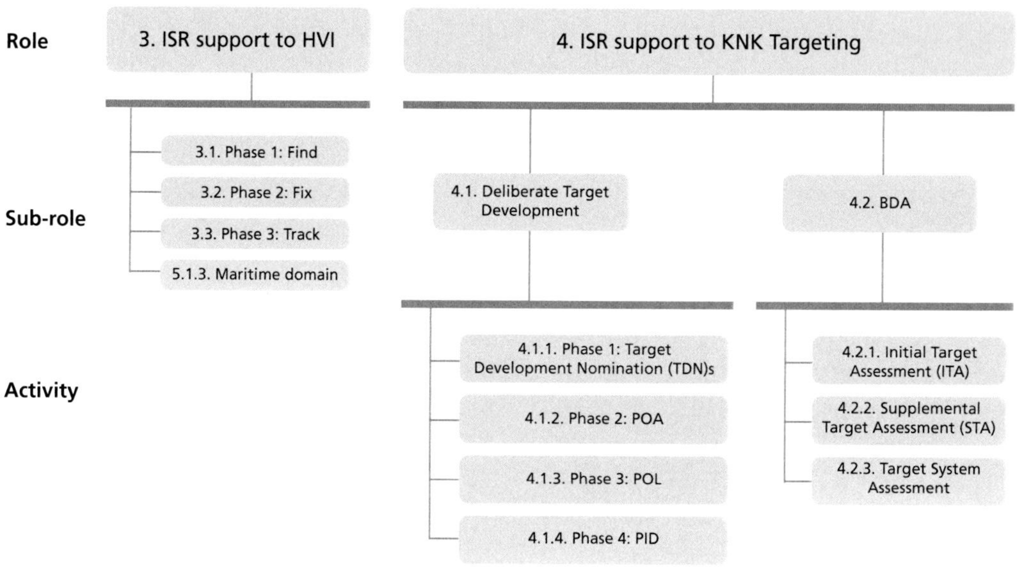

SOURCE: RAND analysis based on CENTCOM input.

1. An ITA provides a physical damage assessment (PDA) and an initial functional damage assessment (FDA) of the target. These assessments are based on multisource intelligence reporting (including ISR assets) available at the time of the assessment.
2. The STA updates the initial report based on availability of additional information. It updates the report on the PDA and the FDA of the target. This report is a detailed PDA, FDA, and change assessment also based on multisource reporting.
3. The TSA aggregates the previous phases of reporting. This assessment is normally produced by national-level intelligence agencies working closely with Command teams but generally does not involve significant additional multidiscipline intelligence.

Strategic Planning

The final category is ISR support to strategic planning. This ISR support provides regular, recurring collection, which contributes to building and maintaining basic INTEL databases that are used for strategic planning. ISR support to INTEL refers to support for the JIPOE, which is the analytic process that joint intelligence organizations use to produce intelligence assessments, estimates, and other intelligence products in support of the joint force commander's decisionmaking process. The operational environment is a composite of the conditions, circumstances, and influences that affect the employment of capabilities and bear on the decisions of the commander.

The JIPOE analyzes all relevant aspects of the adversary, other relevant actors, and the operating environment; the physical domains (air, land, maritime, and space); the information environment, which includes cyberspace; and political, military, economic, social, information, and infrastructure (PMESII) systems and subsystems.[6] JIPOE analyses facilitate the sequencing of intelligence collection requirements and the identification of the most-effective methods of intelligence collection. JIPOE support to ISR is designed to optimize the employment of ISR and target acquisition assets by forecasting the times and locations of anticipated adversary activity. Additionally, ISR collects the information required to update the joint force's JIPOE products. ISR is therefore both a consumer and provider of JIPOE data.[7] This role, sub-roles, and activities are shown in Figure 3.4.

Figure 3.4
ISR Support to Strategic Planning

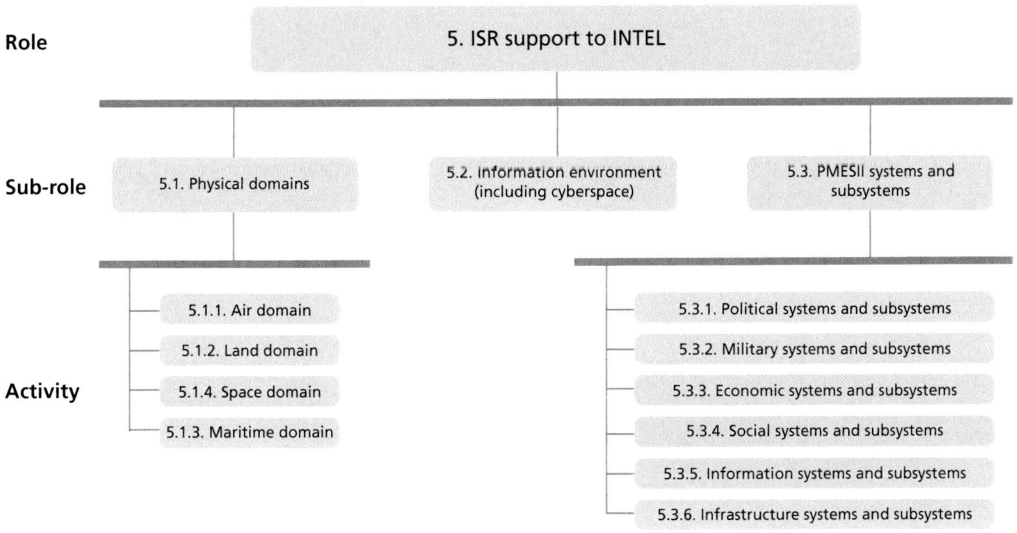

SOURCE: RAND analysis based on CENTCOM input.

6 U.S. Joint Chiefs of Staff, 2014a, page I-1.

7 U.S. Joint Chiefs of Staff, 2014a, p. VII-16.

The roles, sub-roles, and activities described above allowed precise characterization of outputs and outcomes for the key tasks of CENTCOM ISR support. The hierarchy of these tasks proved foundational for the assessment framework, which analyzes the whole through the aggregation of the parts. This work is described in Chapter Four.

CHAPTER FOUR

Assessment Development

Our literature review revealed core principles for ISR tasking and collection, and we used that research to inform the development of MoPs and MoEs for CENTCOM. We are not using MoE and MoP as they are defined in the *DOD Dictionary of Military and Associated Terms* or as used in Joint Publication 5-0 and Chairman of the Joint Chiefs of Staff Instruction 3162.02.[1] Joint Publication 5-0 defines the terms as follows: "measure of effectiveness—An indicator used to measure a current system state, with change indicated by comparing multiple observations over time" and "measure of performance—An indicator used to measure a friendly action that is tied to measuring task accomplishment."[2] In this chapter, we provide our methodological approach for how we developed those MoPs and MoEs, and we provide a framework that could be applied to any CCMD seeking to assess its ISR tasking and collection capabilities. A blank version of the assessment framework is available in Appendix B. Appendix C contains the complete list of the CENTCOM ISR mission, roles, sub-roles, and activities and the metrics (MoEs and MoPs) we developed for assessing these.

Terminology

Specific assumptions underpinned the development of MoEs and MoPs. We have codified those assumptions in the definitions provided below:

- **MoEs** measure *outcomes*. They evaluate how well ISR succeeded in fulfilling the task. Since MoEs focus on the outcome, they should apply to any platform or sensor used.
  - **System MoEs** address the functioning of the collection system or platform itself (e.g., did the sensor collect clear images?).

1 U.S. Department of Defense, *DOD Dictionary of Military and Associated Terms*, Washington, D.C.: U.S. Joint Chiefs of Staff, July 2019; U.S. Joint Chiefs of Staff, 2017a; Chairman of the Joint Chiefs of Staff Instruction, *Methodology for Combat Assessment*, CJCSI 3162.02, March 8, 2019.

2 U.S. Joint Chiefs of Staff, 2017a.

  - **Mission MoEs** address the outcome of the actual collection mission (e.g., did the images provide intelligence value?). It is important to note that we are not assessing the overall value of ISR according to whether a military objective failed or succeeded. The success or failure of an operation depends on many factors, of which ISR is only one. This highlights the difficulty faced by CCMDs in assessing ISR in terms of overall objective accomplishment, when ISR is only one of various factors affecting objective achievement. Instead, our focus is on developing MoEs and MoPs for use in evaluating only intelligence tasking and collection.
- **MoPs** measure *outputs*. They evaluate the performance of the system (e.g., how many images were taken?). Because they focus on particular systems, evaluation of an MoP may vary by platform or sensor used. However, MoPs should be agnostic if the objective of the assessment is to compare systems.
- A **data element** is a variable, such as hours of footage.
- A **sortie** is a single aircraft flight. Data elements for a sortie include flight time, weather conditions, visibility, and other factors.
- A **platform** is the aircraft used to collect ISR data. It can be flown many times a day on many sorties. It can have multiple sensors. Data elements for a platform include fuel level, mechanical condition, configuration, software version, etc.
- A **sensor** is a device that collects intelligence. Each sensor collects many data elements, and different sensors may each collect different data elements, such as hours of footage, number of images, number of targets detected, number of signal intercepts, etc.
- An **observation** is the date-sortie-platform-sensor chain, which will include hierarchical properties (i.e., groups of data elements) relating to the sortie, platform, and sensor.

Metric Development

We developed MoEs and MoPs by combining top-down and bottom-up approaches. The top-down approach started with the assumption that the mission support roles, sub-roles, and activities were defined in joint or service doctrine. A detailed review of these governance documents revealed these definitions. These are the roles, sub-roles, and activities described in Chapter Three. The bottom-up approach assumed that the data already being collected by ISR platforms and sensors (such as the number of sorties flown in a day and the number of images per sortie) measure performance. Our task was to take this quantitative data and develop objective MoEs and MoPs.

The metrics that are easy to measure may not be the metrics that are useful in assessing effectiveness. We combined the top-down and bottom-up approaches as a mechanism to mitigate confirmation bias and anchoring bias:[3] We wanted to know whether the metrics that are easy to collect are actually meaningful, and we did not want CENTCOM's existing metrics to influence our analysis. Therefore, we compared data that are easy for CENTCOM to collect against the roles, sub-roles, and activities in Chapter Three to develop MoEs and MoPs that would be meaningful and useful to CENTCOM decisionmakers.

For our bottom-up approach for each role, we reviewed CENTCOM-developed metrics to determine which role, sub-role, or activity they measure, and we discarded any metric that did not assist in measuring a role, sub-role, or activity and was thus potentially not useful for our purposes. For our top-down approach, we developed MoPs to measure roles, sub-roles, and activities' outputs, and MoEs to evaluate how well the ISR platforms and sensors succeeded in fulfilling the ISR task. For those areas where there was little specific doctrinal guidance or relevant data, we developed new MoEs and MoPs to fill gaps in our lists of MoEs and MoPs.

To assess the ISR tasks unambiguously, we generated MoPs and MoEs that are phrased in the form of a question in past tense to assess the value of ISR activities that have already occurred. Finally, we sought to document the smallest number of MoPs and MoEs that inform the higher-level metric.

The mission MoEs for the ISR roles are shown in Figure 4.1. The MoEs are intended to capture the outcome for the customer of intelligence collection and analysis. These MoEs are fundamental to the assessment of ISR operations—any activity, in fact—because they capture the goal, the standard against which the activities can be measured.

We assessed the sufficiency of available data sources and identified new data required to complete the metrics. We reviewed CENTCOM's current databases and data capture and aggregation processes to ensure that they are synchronized with the MoPs and MoEs we developed, and to ensure consistency in the collection of data that supports long-term analysis. We ensured that metrics are repeatable and could be automated so that data aggregation could occur on any schedule set by CENTCOM. Recurrent aggregation and analysis of data is designed to support consumer demand signal analysis, future trend forecasting, and resource and expertise planning. Analysis of data regularly aggregated using the same collection techniques and same data formats ensures that the data are of high quality and that the resulting analysis of the data is reliable.

To ensure that the MoEs and MoPs were appropriate, we developed a validation process that scored each based on three criteria: reliability, validity, and feasibility. This

---

[3] *Confirmation bias* occurs when an analyst has a preconceived belief in the final answer and intentionally or unintentionally selects data that support this answer while discounting data that would disprove it. *Anchoring bias* occurs when an initial starting point (in this case CENTCOM's existing metrics) creates an anchor point for what "normal" or an expected result should look like. Both of these cognitive biases could prevent analysts from considering or giving equal consideration to all possible solutions.

Figure 4.1
Measures of Effectiveness for ISR Roles

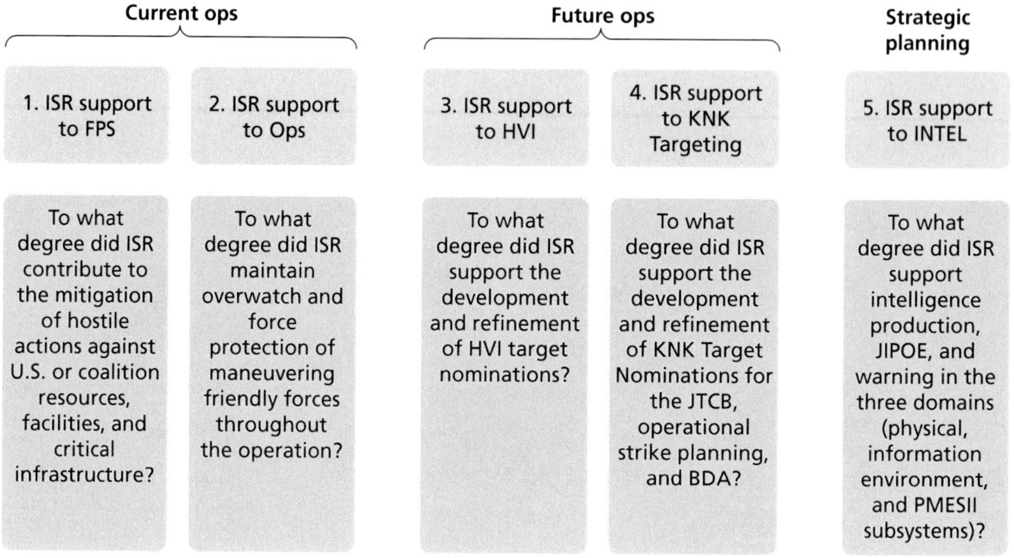

is an internal exercise to test the MoE or MoP itself, not the result. While assessing the reliability, validity, and feasibility would not be used to monitor performance and effectiveness, such an assessment could provide insight into the properties of the metrics themselves.

CENTCOM ISR Assessment Methodology

Once we developed appropriate MoPs and MoEs, we created a framework to apply them to CENTCOM's ISR support mission. The process of collecting raw data elements and applying system MoPs, system MoEs, and mission MoEs is illustrated in Figure 4.2.

The ISR assessment begins with the platforms and sensors that generate raw data, shown on the left of Figure 4.2. ISR platform and sensor operations generate many different kinds of data that are of interest to different stakeholders; these data are used by different stakeholders for many purposes, including performance measurement and assessment.

Those data were used to create ten MoPs, defined in Table 4.1. For purposes of standardization, all of these metrics are computed as percentages ranging from 0 to 100 percent. These metrics were then binned into three categories of MoPs, according to whether they inform ISR tasking, ISR collection, or the aircraft.

Figure 4.2
Metrics Breakdown

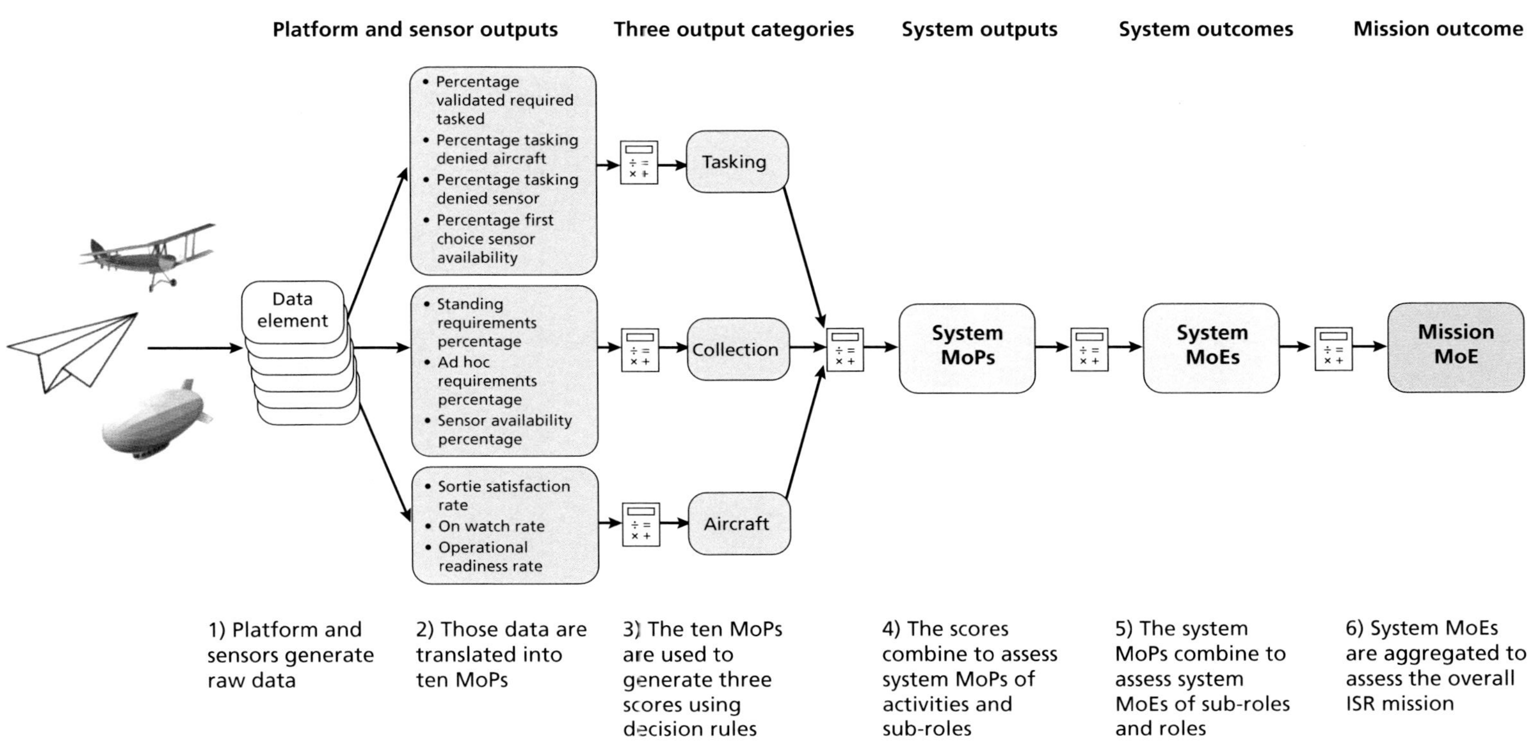

Table 4.1
ISR Assessment Metrics

Output category	MoP Name	Definition
Aircraft	Sortie satisfaction	Number of sorties flown/total number of sorties scheduled
	On watch	Number of hours of planned on watch, per sortie/number executed hours
	Operational readiness	Total number of hours flown/total number of hours planned
Collection	Standing requirements satisfied	Number of standing requirements collected/ number of standing requirements tasked
	Ad hoc requirements satisfied	Number of ad hoc requirements collected/ number of ad hoc requirements tasked
	Sensor operational	Number of hours sensor was operational/ number of planned hours
Tasking[a]	Validated requirements denied	Number of taskings denied/total number of taskings requested
	Tasking denied because platform unavailable	Number of taskings denied because of lack of available platform/total number of taskings requested
	Tasking denied because sensor unavailable	Number of taskings denied because of lack of available sensor/total number of taskings requested
	First-choice collection sensor unavailability	Number of times sensor was unavailable/ number of times the sensor was requested

NOTE: [a] We understand that, as drafted, such metrics as "first-choice sensor unavailability" are not supportable by the data the CENTCOM J2 currently collects. We suggest that collecting these or similar data, however, is necessary to determine tasking performance and effectiveness.

The ten MoPs were selected by our team because they are commonly collected by ISR platforms and sensors, and their data are collected over a long period in a standardized format, meaning the data are generally reliable. We developed three bins for the ten elements to facilitate analysis of different activities during different phases of an ISR mission.

To generate higher-level MoPs that measure performance for ISR tasking, ISR collection, and the aircraft, we aggregated the metrics within each output category to generate "scores." This is accomplished by first establishing and then evaluating decision rules. For simplicity, our scores are evaluated along an ordinal scale of red, yellow, or green. An example of the decision rule for the Aircraft category might be

- *Green* if each of the three composite metrics has an average above 80 percent,[4]
- *Red* if any of the three composite metrics has an average below 60 percent
- *Yellow* otherwise (at least one of the composite metrics has an average that is less than 80 percent, but the averages of all three are each greater than 60 percent).

With this set of decision rules, a sortie with a sortie satisfaction rate of 73 percent, an on watch rate of 67 percent, and an operational readiness rate of 92 percent would result in a yellow Aircraft rating.

The Tasking and Collection categories could be computed similarly. An alternative (yet more complicated) set of decision rules could be developed using a similar scorecard found in Joint Publication 2-01, *Joint and National Intelligence Support to Military Operations* (see Appendix A, Table A.2), which defines a five-point Likert scale that ranges from "completely ineffective" to "completely effective." However, the construction of appropriate decision rules is a subjective task, with no single or correct formulation.

Moving on to the system MoPs and system MoEs, each activity, sub-role, and role will be evaluated in a similar method. Decision rules will supply ratings at each level and, ultimately, for the ISR mission overall.

For example, ISR support to ops contains two sub-roles (overwatch and dynamic targeting), and overwatch consists of four activities (high-value unit, non–high-value unit, U.S. forces, and American citizens). Each of the four activities is evaluated separately using the ten metrics and three scores (i.e., each activity receives a "Red," "Yellow," or "Green," based on the decision rules applied to the component scores for Tasking, Collection, and Aircraft). Then, these four activities are evaluated together to assess the system MoP for the overwatch sub-role according to another predetermined decision rule. (For example, the results of the four activities could be averaged together, the lowest value could become the overall rating if a failure in one area is a failure for the entire mission, or another decision rule could be created.) Overwatch and dynamic targeting are then combined together with a third decision rule to assess the system MoE for the ISR support to ops role. System MoEs for the roles are then aggregated to generate the mission MoE to assess the overall ISR support mission.

We produced a blank framework of this assessment as Appendix B.

---

4 These ranges are notional, and CENTCOM can decide the appropriate range it uses to indicate high, medium, and low.

CHAPTER FIVE

Data Visualization

The assessment described in Chapter Four may be conducted directly in Excel. The use of data visualization software, however, could facilitate additional analysis and allow the user to customize views of the assessment results. Therefore, we produced a visualization tool using Tableau software that may be used to conduct the assessment and display the results.

The visualization tool presents the results of the ISR assessment process in a format that is both easily digestible at first glance and also conducive to deeper investigation into the findings. In building the tool, we considered the needs of such users as senior intelligence and ops officers. Although the assessment itself provides MoPs; MoEs; and scores for the roles, sub-roles, and activities of CENTCOM ISR support, much more information could be obtained from additional data correlations within that assessment. The various views of the data, known as "dashboards," provide those correlations and answer a broad range of questions: In what circumstances does a specific platform fail? How did a given sub-role perform during February 2018? What is the status of the ISR mission overall?

To make this report and the tool itself accessible to as wide an audience as possible, the visualization tool described in this chapter uses data that are inspired by the CENTCOM Office of ISR Assessments database for the data visualization. The methodology for the tool, which includes detailed instructions for using and reconstructing the tool, is provided separately.[1]

Visualizing MoPs and MoEs

To visualize the various aspects of the CENTCOM ISR mission, we constructed five distinct and complementary dashboards, two of which are reproduced next.

[1] Amanda Wicker, Sasha Romanosky, Cortney Weinbaum, Bradley Knopp, and David Luckey, *Measuring Intelligence, Surveillance, and Reconnaissance Effectiveness at the United States Central Command: Data Visualization Tool Documentation*, Santa Monica, Calif.: RAND Corporation, TL-358-OSD, 2020.

(The entire set of dashboards is included in Appendix D.) The first dashboard, in Figure 5.1, shows an overview of the status of the CENTCOM ISR mission. In this figure and all other dashboards, all platform, sensor, and operation names are fictional.

From this starting point, the user can click down into more granular information, such as the performance of sub-roles and activities. The view of the ISR mission can also be refined through various filters: system MoP, platform, sensor, operation, country, region, and date range.

At the bottom of the *Overview Dashboard*, there are charts that show the system scores that roll up into the system MoP, the system MoP by platform and by sensor, and the platform and sensor MoPs that roll up into the Aircraft, Collection, and Tasking scores. Clicking on a square (representing a role, sub-role, or activity) filters the values in the charts (Scores, System MoP by Platform and by Sensor, and MoPs). This step-by-step process for one sub-role is annotated in Figure 5.2.

This design allows a quick investigation of ISR mission status to enable decision-making. If a role is not performing well, the data responsible can be quickly revealed by examining the visualization. For example, an investigation of the support to the FPS role reveals that sub-role 1.2 (small tactical units) is the source of low scores. The

Figure 5.1
Overview Dashboard

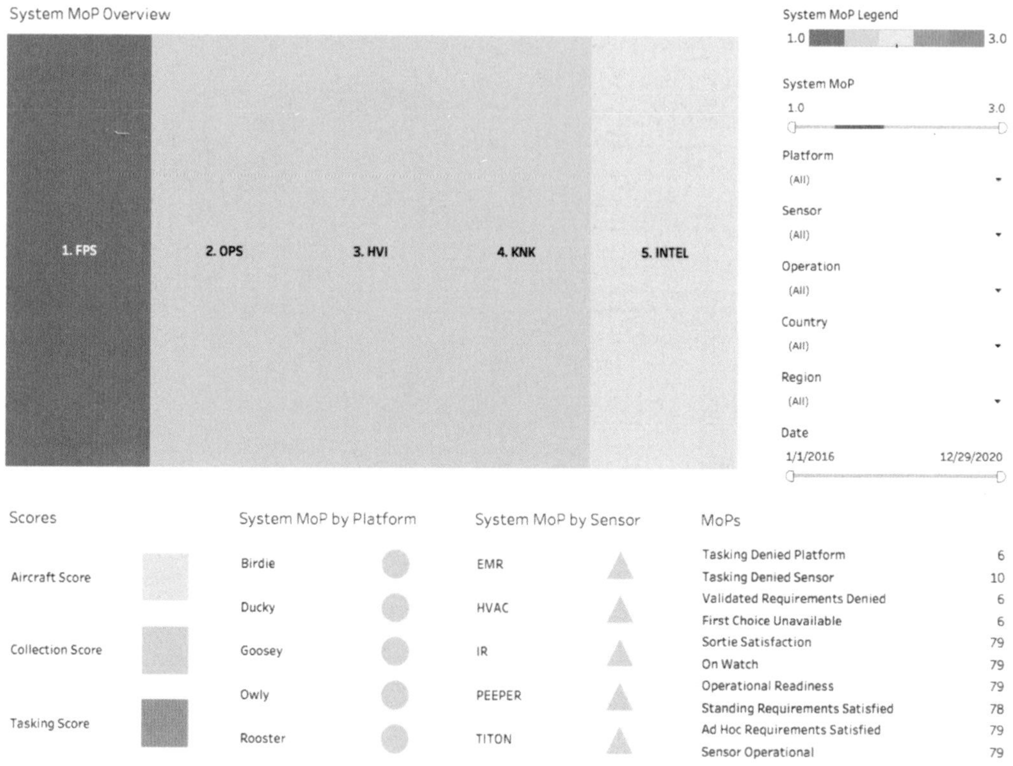

Figure 5.2
Annotated Overview Dashboard Filtered for One Sub-role

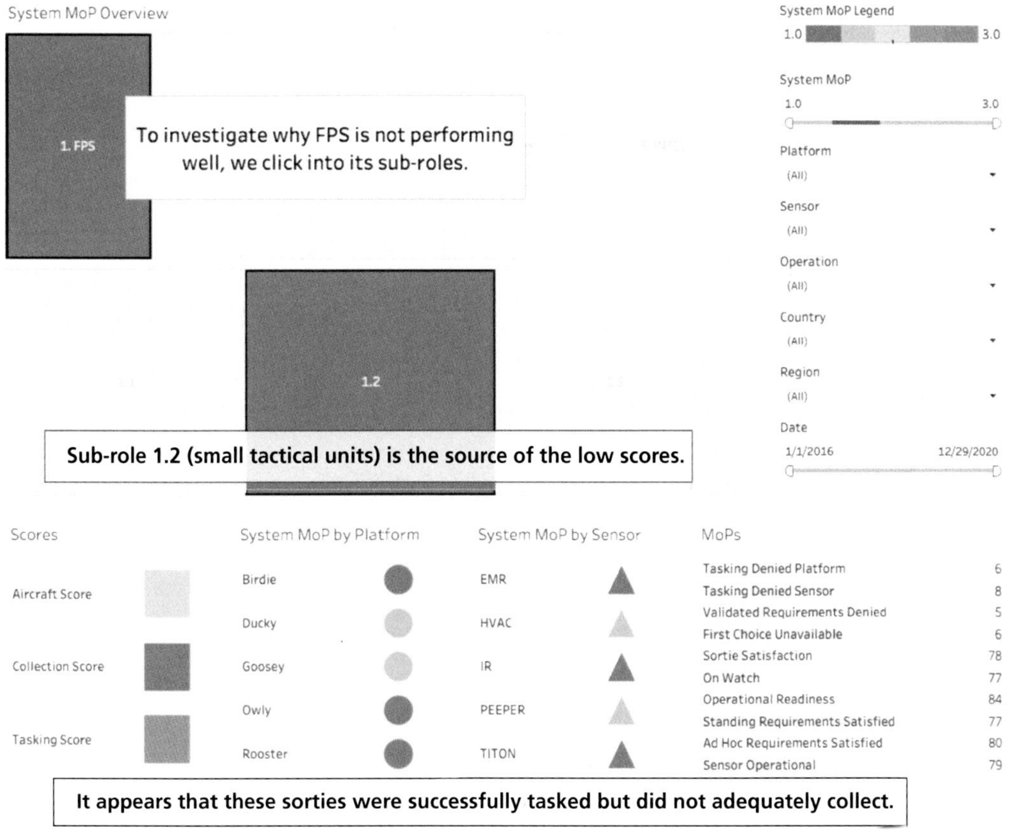

data show that the Tasking score was high, but the Collection score was poor, which might suggest that the sensor was not operational or that collection was failing to meet requirements. Here, the analyst could turn to more-tailored data sources to investigate the specific problem.

Although the Overview Dashboard allows for a considerable range of analysis, we also constructed other visualizations to respond to more specific needs, and these are provided in Appendix D. In that section, the *Map Dashboard* allows the user to see performance by region and by operation. The side-by-side views in the *Platform* and *Sensor Comparison Dashboards* allow the user to directly compare two different sensors or platforms under a variety of specific conditions. The *Missing Data Dashboard* is a solution to CENTCOM's concern that the sorties that do not generate any data are not well tracked. These additional visualizations are in Appendix D.

The different dashboards that we developed for this tool can supply a richer understanding of the status of the CENTCOM ISR mission by answering key questions in easily interpreted data visualizations. Although this tool was designed to visualize ISR support to CENTCOM data, it could be adapted to depict performance for a wide variety of organizations.

Limitations

Because of the sensitive nature of the ISR data and our lack of access to all relevant fields of data we needed, we developed the visualization tool using simulated data modeled off the kinds of data CENTCOM would receive from its platforms and sensors.

This tool does not weight the relative importance of activities to a sub-role or of sub-roles to a role, and a single set of decision rules is applied for all levels. For example, are high-value unit, non–high-value unit, U.S. forces, and American citizens all equivalent in measuring the success of overwatch? Are overwatch and dynamic targeting equally important to achieving ISR support to ops? Should the threshold for on watch rate be different for INTEL and FPS? These questions were not addressed in the scope of this project but should be carefully considered by CCMDs or others who wish to incorporate our assessment methodology, as described in Chapter Three, into their own practices. Instead of including subjective weightings or varied decision rules, this sample visualization is constructed with all metrics rolled up to higher levels as averages. This may not reflect real-world situations in which one task might be critical to success of the mission, while others are merely contributory.

An additional limitation that results from applying the same decision rule to all levels is that the tool combines scores to assess system MoPs of activities, sub-roles, and roles. This differs from the approach described in Figure 4.2, where a system MoP is generated with scores from activities and sub-roles, and these system MoPs, in turn, are combined to assess system MoEs for sub-roles and roles. In our tool, neither a system MoE nor a mission MoE is generated.

Another simplification we have made for this visualization is that synthetic data have only one platform and one sensor per sortie. In practice, multiple types of sensors on a single platform may contribute data for a single sortie. These sensors may provide different types of intelligence, or different amounts or quality of data. The relative value of these different sensor data, in conjunction with the platform and even the mission, must be determined to properly visualize their performance and effectiveness. Future analysis could serve to address these limitations in the tool.

CHAPTER SIX

Observations, Findings, and Conclusions

Many analyses of ISR effectiveness have been conducted in the past to support operational and resource planning, but, based on our review, few of them examined the data derived from operations to improve the understanding of ISR output and outcomes. To broaden the comprehensiveness of our study, we examined the public, academic, and private sector literature, specifically looking into the development of performance metrics. Our team examined how other organizations develop and apply performance and effectiveness metrics, and then we applied those lessons to CENTCOM's particular circumstances.

Performance effectiveness assessments are, by definition, closely related to the organization performing the operations. Therefore, the metrics used to conduct the assessments need to be tailored to the user and defined by user expectations or outcome. For this study, we made assumptions that allowed us to focus on CENTCOM-specific issues. We used Command-defined objectives to develop MoEs and MoPs that were analyzed to gauge operational effectiveness. While the specific metrics were developed for CENTCOM, the methodology is applicable to other organizations.

The most pertinent finding across literature for CENTCOM relates to MoPs and MoEs—specifically, that MoPs must focus on outputs (usually a quantitative measure, or binary), while MoEs need to measure mission outcomes (usually qualitative). Both sets of metrics must be defined clearly in the context of their use and should be scrutinized to ensure that they can be gathered as needed (reliability measure), accurate (validation measure), and feasible (ensures that the metric design focuses on data and systems currently in use).

Determining how to measure performance is complex, and, generally, more stakeholders lead to more-complex assessments. Different stakeholders have different needs from the assessments and often want to use different data for their analyses. Previous RAND studies of organizations that use extant data for analysis have revealed that these data are often less complete and less reliable than data owners realize. In these cases, data heritage and data reliability become linchpins in the analysis and are often cited by critics as a basis for devaluing the analysis and the assessment.

Throughout this study, the RAND team and CENTCOM considered data and databases that were needed for performance assessment evaluations. We uncovered several issues that were outside the scope of our study regarding data. Many of these data issues could be uncovered in other studies, and we found them manifested as follows:

- Data heritage: Merging data from different sources using different formats and taxonomies leads to questions about the accuracy of the data that reduce confidence in the output from data analytics. Fragmented ownership of the data means the data quality and standards are driven by multiple stakeholders. The absence of data collection and aggregation standards should be addressed early to minimize these concerns.
- Data curation: A major frustration for assessment staffs is the amount of time and resources it takes to prepare data for use: cleaning, sorting, scrubbing, and deduplicating before the data can be accurately recorded and used. Data integrity becomes an issue in the curation process; users are consistently concerned about making inadvertent changes during the curation process and potentially introducing errors into the data set.
- Data volume: For data collectors and aggregators, data maintenance could be one of the major difficulties. The large size of databases and the complexity of database architectures mean they are not easy to maintain. For data users, volume raises the question of the need for tools to exploit the data. Different stakeholders often need different tools to accomplish their work, but many users complain that they cannot find organization "advocates" who can make the case for acquiring and maintaining exploitation tools.

A basic assumption for this research is that we are attempting to measure the inherently unmeasurable by seeking to identify quantitative metrics as proxies for otherwise difficult-to-measure things. Selection of proxy metrics needs to be meticulously done so that the metrics, in aggregate, address the question at hand. Each proxy measure, however, needs to be assessed individually to ascertain its reliability as a measure, the feasibility of collecting the data for the metric, and the validity of the data once collected. These three characteristics are averaged to arrive at a characterization of the metric and, thus, an assessment of its utility to the assessment process. For our analysis, this process was conducted by identifying the effect CENTCOM ISR ops aim to achieve for customers and other stakeholders and translating that into MoEs and MoPs to allow more detailed and quantitative analysis.

Finally, our team concluded that using a visualization tool to display the assessment results allows analysts and other stakeholders to use the data to enable decisions within their realm of authority. Such a tool allows users to alter the parameters of the data for purposes of analysis and to see correlations that would not otherwise be easily visible. Map-based visualization also allows users to see performance and effectiveness by region and operation, information that enhances decisionmaking.

The development of repeatable, scalable, data-informed metrics can assist with improving the quality, effectiveness, and efficiency of ISR assessments. While this project specifically analyzed the outputs and outcomes of CENTCOM ISR support ops, the methodology and findings could also be applied to organizations and stakeholders inside and outside the federal government that conduct intelligence activities of any type or any organization that assesses mission performance or effectiveness.

APPENDIX A

Literature Review

Overview

CENTCOM's Office of ISR Assessments asked RAND to develop a methodology to evaluate CENTCOM's ISR activities by measuring tasking and collection operations within its AOR. We scoped our review to an assessment of ISR platforms and sensors, not to how such data are exploited or incorporated into finished intelligence products.[1] We reviewed public sector and private industry performance assessment practices, and we established baseline government policies for ISR processes. This literature review is relevant to readers who desire an increased awareness of ISR-metric practices in the wider ISR community. This review is organized according to the themes that emerged from academic, private sector, and government literature focused on ISR assessments.

Methodology

The RAND team sought literature from across the three domains of government, academia, and private sector publications. The team sought literature that was developed over the previous decade to ensure (1) that we understood legacy ISR assessment approaches, and (2) that findings from the literature remained relevant to ISR assessment practitioners. Research of government documents included official DoD directives and instructions on ISR and, specifically, the conduct of ISR assessments. Our research also included examination of joint-level and CCMD guidance intelligence doctrine of the Joint Staff and services, but the contents of such doctrinal publications did not mention ISR assessments; as a result, they were excluded from this review. Lastly, we excluded publications that focus only on the PED phases of the TCPED process, which are beyond the scope of this research, but we acknowledge that there is a range of literature dedicated to this topic.[2] The sections of this literature review are

---

[1] The full spectrum of ISR support includes find, fix, finish, exploit, analyze, and disseminate, also known as the *F3EAD* process in DoD parlance.

[2] For example, RAND researchers have performed in-depth studies on U.S. military Motion Imagery Processing and Exploitation processes: Amado Cordova, Lindsay D. Millard, Lance Menthe, Robert A. Guffey, and Carl Rhodes, *Motion Imagery Processing and Exploitation (MIPE)*, Santa Monica, Calif.: RAND Corporation, RR-

arranged in the following manner. First, we present the mission of ISR according to existing doctrine. This is followed by the general ISR assessment information we collected from DoD sources. In closing, this review reveals three major themes evidenced by the literature that could assist in future ISR assessment planning: (1) aligning program requirements and assessments, (2) data latency, and (3) generalizing and consolidating metrics to increase assessment utility, before proceeding to overall conclusions and recommendations.

What Is the Mission of Intelligence, Surveillance, and Reconnaissance?

ISR definitions and objectives vary between sectors and are context dependent. DoD defines ISR as

> 1. An integrated operations and intelligence activity that synchronizes and integrates the planning and operation of sensors, assets, and processing, exploitation, and dissemination systems in direct support of current and future operations. 2. The organizations or assets conducting such activities.[3]

Service component definitions usually represent variants of this overarching definition. For example, the U.S. Air Force (USAF) uses a different term to represent integration of ISR functions, calling it *Global Integrated ISR*, defined as the

> cross-domain synchronization and integration of the planning and operation of ISR assets; sensors; processing, exploitation and dissemination systems; and, analysis and production capabilities across the globe to enable current and future operations.[4]

DoD ISR capabilities and functions can also support others within the defense and intelligence community. As DoD has sought to refine ISR capability requirements and integrate with the wider defense community, the department has faced numerous challenges. For example, a 2008 U.S. Government Accountability Office (GAO) report notes that the DoD was

---

154-AF, 2013; Amado Cordova, Kirsten M. Keller, Lance Menthe, and Carl Rhodes, *Virtual Collaboration for a Distributed Enterprise*, Santa Monica, Calif.: RAND Corporation, RR-153-AF, 2013; Lance Menthe, Amado Cordova, Carl Rhodes, Rachel Costello, and Jeffrey Sullivan, *The Future of Air Force Motion Imagery Exploitation: Lessons from the Commercial World*, Santa Monica, Calif.: RAND Corporation, TR-1133-AF, 2012.

3 U.S. Joint Chiefs of Staff, 2017c; DoD, 2019.

4 USAF Annex 2-0, *Global Integrated Intelligence, Surveillance & Reconnaissance Operations*, Montgomery, Ala.: Curtis E. Lemay Center, January 29, 2015.

> faced with different and sometimes competing organizational cultures, funding arrangements, and requirements processes . . . this wide range of DoD ISR enterprise commitments across the U.S. intelligence community presents challenges for DoD as it works to increase ISR effectiveness and avoid unnecessary investments in ISR capabilities.[5]

The deployment of ISR capabilities within a CCMD AOR is used to achieve a wide variety of theater effects. For example, ISR might be used to track insurgent activity, provide situational awareness to maneuvering ground forces, or—in CENTCOM's case—to identify precursor materials for the manufacture of improvised explosive devices (IEDs).[6] The assessment of the tasking and collection stages are crucial for future CCMD investment decisions; not only does assessment at this stage provide critical answers to questions such as "*are we using the right platform for the right job?*" but also to whether a commander or CCMD's requirements could be fulfilled with particular sensors deployed on such a platform.[7]

JIPOE provides another doctrinal pillar that feeds the ISR mission. The JIPOE is used by joint commands, agencies, and services to produce intelligence products that support commander decisionmaking. While the JIPOE mainly serves to inform CCMDs about adversary "capabilities and intentions" in the context of the operational environment, ISR missions collect needed information to update JIPOE products. The 2014 JIPOE suggests that ISR is "therefore both a consumer and provider of JIPOE data."[8]

General Guidance on ISR Assessments from DoD Sources

There are a variety of government sources that offer ISR assessment guidance for J2 mission planners, although they do not always offer the level of granularity required for an effective assessment framework. Two particular government sources, however, could serve as a basis for building such a metrics framework: the 2017 *Joint and*

---

5 GAO, *Intelligence, Surveillance, and Reconnaissance: DOD Can Better Assess and Integrate ISR Capabilities and Oversee Development of Future ISR Requirements*, GAO-08-374, Washington, D.C., March 2008; GAO, *Intelligence, Surveillance, and Reconnaissance: Actions Are Needed to Increase Integration and Efficiencies of DOD's ISR Enterprise*, GAO-11-465, Washington, D.C., June 2011b.

6 This requirement (Counter-IED) may not exist across other CCMDs and is purely illustrative.

7 ISR missions are also conducted under the auspices of the F3EAD intelligence-fusion cycle. Such missions are normally carried out in support of unique special ops targeting missions and account for the ongoing activities known as *targeting*.

8 U.S. Joint Chiefs of Staff, 2014a.

National Intelligence Support to Military Operations guidance (Joint Publication 2-01) and the 2011 *Commander's Handbook for Persistent Surveillance*.[9]

Joint Publication 2-01 explains that the purpose of ISR assessments is to "evaluate the performance of intelligence collection operations in order to improve collection effectiveness in meeting intelligence and operational requirements."[10] The Joint Staff suggests that ISR assessments drive a "continual improvement process through all phases of the intelligence process by identifying actionable recommendations to influence the ISR strategy, as well as collection asset/resource allocation and employment," and that "continuous and timely assessment is crucial to monitor and measure progress toward mission accomplishment."[11] Joint Publication 2-01 lists four critical items an ISR assessment should measure: (1) the value of intelligence gain; (2) the reasons why ISR activities were or were not successful in answering the intelligence problem; (3) how well the intelligence problem was answered, and (4) what actions need to be taken to address poor performance or limited effectiveness.[12] Joint Publication 2-01 ultimately suggests that the goal of ISR assessments is to "provide lessons learned that can improve ISR performance and avoid waste of resources, time, and PED capacity."[13]

To answer these four questions, Joint Publication 2-01 suggests the use of quantitative performance measures to address whether the ISR capability performed "within technical standards and whether the planned collection was accomplished," whereas qualitative effectiveness measures would answer whether the "collection that was accomplished satisfied the requirement."[14] Table A.1 depicts these and other important definitions gleaned from Joint Publication 2-01.

Joint Publication 2-01 also suggests the use of numerical scoring to compare ISR performance, i.e., MoPs, to determine whether "steps to improve performance are moving in a positive or negative direction."[15] The numerical scoring schema would occur as part of a three-step process: (1) scoring MoPs within bound of percentages (see Table A.2); (2) scoring MoEs along an effective gradient, e.g., "ineffective" to "completely effective"; and (3) placing the numerical score gleaned from observations of the platform and sensor into the context of the overall "intelligence problem."[16]

---

9 Although U.S. Joint Forces Command no longer exists, the 2011 *Commander's Handbook* contained pertinent information to help our understanding of ISR assessment functions. See *Commander's Handbook for Persistent Surveillance*, 2011, p. I-1.

10 U.S. Joint Chiefs of Staff, 2017c.

11 U.S. Joint Chiefs of Staff, 2017c.

12 U.S. Joint Chiefs of Staff, 2017c, p. B-7.

13 U.S. Joint Chiefs of Staff, 2017c, p. B-11.

14 U.S. Joint Chiefs of Staff, 2017c, p. B-7.

15 U.S. Joint Chiefs of Staff, 2017c, pp. B-10 and B-11.

16 U.S. Joint Chiefs of Staff, 2017c, pp. B-10 and B-11.

Table A.1
Joint Publication 2-01 ISR Guidance and Key Definitions

Key Term	Definition and Examples
Assessments measures	Measure performance, efficiency, and effectiveness
MoPs	Measure task accomplishment by evaluating whether the ISR activities met a measurable standard; give indication of the extent of progress in execution of the plan; should be generally focused on the immediate results of tactical actions
MoEs	Help determine how well the mission is being accomplished; involve a component of subjective evaluation on the basis of objective data; should be based on observable and measurable indicators
Formative assessments[a]	Determine how well an ISR activity was performed and what can be done to improve the next mission; occur after each mission to fine-tune requirements for the next mission: e.g., Did the ISR capabilities perform within technical standards? What was the volume of requirements collected during the mission? Were mission objectives met? How did this mission contribute to answering the intelligence problem? Do the mission objectives need to be adjusted for the next mission?
Summative assessments[b]	Calculate overall contribution of ISR activities in meeting mission objectives and answering intelligence problems during a specified period: e.g., How well were mission objectives met? How well did ISR activities answer intelligence problems? To what extent was the right mix of ISR capabilities employed? To what extent were ISR assets available to accomplish missions? What are recommended improvements for performance, efficiency and effectiveness in accomplishing mission objectives?

SOURCE: U.S. Joint Chiefs of Staff, 2017a.

NOTES: [a] We suggest that the definitions and examples provided in the formative assessments section are apt descriptions of the assessments conducted by the CENTCOM J2 and the types of questions they attempt to answer.

[b] We suggest that the definitions and examples provided in the summative assessments section are apt descriptions of the assessments conducted by the CENTCOM J2 and the types of questions they attempt to answer.

Chapter Six of the *Commander's Handbook for Persistent Surveillance* offers four additional considerations related to ISR assessment:[17] (1) Was sufficient planning accomplished that focused on achieving the collection outcome? (2) Were the collection assets tasked appropriately for the collection requirement? (3) Was sufficient time allocated for the collection requirement to ensure success? (4) Was the information collected within the time required? These all serve as important considerations in assessing the tasking and collection cycle for CENTCOM ISR planner implementation.[18]

[17] *Commander's Handbook for Persistent Surveillance*, 2011b.

[18] Definitions for measures of performance and effectiveness are largely unchanged in Joint Publication 2-01 from the 2011 *Commander's Handbook for Persistent Surveillance*. For example, the handbook states that an MoP is "used to measure accomplishment of the ISR task." MoPs should answer whether the collection mission took place, whether it obtained the essential elements of information (EEIs) linked to the collection requirement, and

Table A.2
Performance Assessments Scorecard Example

Performance Assessments Scorecard						
MoP	**Not Assessed**	**Completely Ineffective**	**Mostly Ineffective**	**Somewhat Effective**	**Mostly Effective**	**Completely Effective**
Performance: At what quantity level is this intelligence/ capability fulfilling tasking?	not applicable	Asset is performing at an unsatisfactory level or way below its technical specifications; meets less than 20% of collection hours compared with tasked hours.	Asset is performing subpar and is not meeting expected capabilities; meets between 20–39% of expected collection hours compared with tasked hours.	Asset is performing adequately with only minor issues that hamper performance; meets between 40–59% of expected collection hours compared with task hours.	Asset is performing at near optimal level; meets between 60–79% of expected collection hours compared with task hours.	Asset is performing flawlessly; meets more than 80% of expected collection hours compared with task hours.
Numerical score	not applicable	0–19%	20–39%	40–59%	60–79%	80–100%

SOURCE: U.S. Joint Chiefs of Staff, 2017c, p. B-11.

Obstacles to Effective ISR Assessment

Aligning Program Requirements and Assessments

ISR assessment practitioners across CCMDs tend to focus on the development of quantitative MoPs complicating a logical linkage to qualitative MoEs. This assessment process is further complicated through the alignment of unique CCMD priority intelligence requirements (PIRs) to various ISR functions.[19]

A previous RAND report for the USAF identified similar CCMD challenges, its authors observing that "the majority of ISR assessments have focused on using statistics from the tactical level (e.g., sorties flown and percentage of planned images collected)"

whether it gathered the desired information. They are used to evaluate the function of an ISR platform and to answer the "what," "where," and "when" questions for a specific collection requirement. Mission performance can be measured by the quantity of intelligence products generated, the number of collections tasked to multiple assets and the complexity of the mission, and the percentage of missions that have been affected by internal and external problems and therefore might require retasking. The 2011 *Commander's Handbook for Persistent Surveillance* defines MoEs as assessing "whether a collection mission sufficiently answered the essential elements of information of a collection requirement." Ideally, MoEs should assess whether a collection mission was able to provide useful information about either the adversary or the operational environment, whether the intelligence requirements were satisfied, and whether the intelligence collected supported the decisionmaking process. The MoEs are determined by addressing the question of whether the information gained from the mission allowed the commander to make a timely decision. Mission effectiveness can be measured by whether the persistent surveillance mission achieved the persistent surveillance objective, and whether the persistent surveillance mission supported the operational objectives. See *Commander's Handbook for Persistent Surveillance*, 2011b, pp. VI-1 and VI-3.

[19] This section notes the importance of having metrics aligned to task objectives; accordingly, it focuses primarily on DoD references. The RAND team did not identify information that was relevant to this category within private sector or academic literature.

but leaving out whether the ISR system is "satisfying the commander's intent" at the operational and strategic level.[20] The report concluded that CCMDs should adopt a "strategy-to-task" framework to enable end-to-end assessment processes for daily operations.[21] There are some aspects to measuring effectiveness that involve more quantitative assessments than notional or qualitative ones. That RAND report also examined methodologies for improving ISR assessments and notes that the ISR collection and processing step contains mostly quantitative measures, although it sometimes includes "collection satisfaction" related indicators—such as "did the customer get the image they needed?"—that could produce lower success rates than anticipated.[22] The challenge of obtaining high-quality, informative customer feedback in response to these questions, therefore, presents ISR managers and developers with the added task of soliciting, maintaining, and evaluating customer responses.[23]

No single definition for airborne ISR capabilities or requirements exists throughout the literature we examined.[24] Rather, many ISR definitions are meant to assist global force management in matching ISR platforms to specific CCMD missions. Joint Publication 2-01 explains that each ISR capability requirement should be "realistically defined in a manner and with sufficient fidelity to allow strategic ISR managers to allocate resources and assets to meet the requirement."[25] Not having a single definition for ISR may allow a greater degree of flexibility across CCMDs, although a narrow focus on matching platforms to tasks may lose important nuance on the actual effectiveness of collection operations.

---

20 Sherrill Lingel, Carl Rhodes, Amado Cordova, Jeff Hagen, Joel Kvitky, and Lance Menthe, *Methodology for Improving the Planning, Execution, and Assessment of Intelligence, Surveillance, and Reconnaissance Operation*, Santa Monica, Calif.: RAND Corporation, RB-242-AF, 2008, p. xii.

21 Lingel et al., 2008, p. 36.

22 Lingel et al., 2008, p. 39.

23 Lingel et al., 2008.

24 In addition, we find that Chairman of The Joint Chiefs of Staff Manual (CJCSM) 3314.01A and Chairman of the Joint Chiefs of Staff Instruction (CJCSI) 3340.02B contain two different definitions of ISR. CJCSM 3314.01A, citing Joint Publication 2-0 as a reference, says that ISR is "An integrated operations intelligence activity that synchronizes and integrates the planning and operation of sensors, assets, and the processing, exploitation, and dissemination systems in direct support of current and future operations. This is an integrated intelligence and operations function," while CJCSM 3340.02B cites Joint Publication 2-01 (U.S. Joint Chiefs of Staff, 2017c) as a reference, stating that ISR is "An activity that synchronizes and integrates the planning and operation of sensors, assets, and processing, exploitation, and dissemination systems in direct support of current and future operations." Chairman of the Joint Chiefs of Staff Manual, *Intelligence Planning*, CJCSM 3314.01A, September 17, 2012; Chairman of the Joint Chiefs of Staff Instruction, *Joint Enterprise Integration of Warfighter Intelligence*, CJCSI 3340.02B, October 24, 2013.

25 Joint Publication 2-01 also states that "CCMD ISR planners should apply a systematic process to ensure a coherent linkage of PIRs/collection requirements to airborne ISR capability requirements is made." See U.S. Joint Chiefs of Staff, 2017c, p. B-2.

Brig Gen Timothy D. Haugh and Lt Col Douglas W. Leonard provide observations to assist in the framing of ISR-assessment capabilities: First, "Did services acquire the right ISR capabilities in the right number, performing as designed?" Second, "Were the available theater airborne ISR capabilities apportioned correctly?" And third, "Was the theater airborne ISR employed effectively?"[26]

Haugh and Leonard note that some of the most-common metrics for assessment within the ISR community include the number of ISR sorties planned and executed, sensor availability, the number of images collected, the number of EEIs satisfied, the number of hours of full-motion video products produced, and the number of intelligence products produced.[27] They also note, however, that these types of indicators do not help ISR managers assess overall ISR effectiveness (e.g., did ISR advance the supported commander's desired operational outcomes? or, did the use of ISR close intelligence gaps?).[28] This lack of a requirement-to-strategy linkage is further expounded upon in the following observation they make:

> [T]he CFACC's intelligence team developed a separate reporting mechanism to track the thousands of intelligence reports provided to coalition partners and reported these results to CENTCOM and OSD monthly, though that mechanism included only raw numbers without an effort to link those specific products back to supported outcomes or gaps.[29]

They conclude that ISR platforms are rarely evaluated based on the ability to produce intelligence that closes intelligence gaps; therefore, current ISR assessments do not meaningfully contribute to CCMD decisionmaking nor increase a collective understanding of the operational environment.[30] See Figure A.1 for information about the disconnect between MoEs and MoPs for ISR.

- **Finding:** Higher-level assessment strategies require inputs from multiple stakeholders at subordinate levels to ensure that a logical linkage between MoPs and

---

[26] Timothy D. Haugh and Douglas W. Leonard, "Improving Outcomes: Intelligence, Surveillance, and Reconnaissance Assessment," *Air & Space Power Journal*, Winter 2017.

[27] Haugh and Leonard, 2017.

[28] Haugh draws from Maj Gen. Jack Shanahan's notion of the "three rights" to develop a proposed framework. Shanahan's three rights were "Right Intelligence, Right Person, Right Time: Delivering the right ISR to the right person at the right time" to turn "data into information, information into knowledge and knowledge into actionable intelligence that results in better decisions." See Haugh and Leonard, 2017, p. 7.

[29] Haugh and Leonard, 2017, p. 11.

[30] Establishing intelligence priorities and properly tasking assets is an important step of ISR collection. Commanders use commander's critical information requirements (CCIRs) to prioritize collection assets and ensure efficiency. In this way, CCIRs allow commanders to ensure that collection assets are properly apportioned to support their intelligence requirements. See, for example, Deployable Training Division, *Intelligence Operations*, first edition, Insights and Best Practices Focus Paper, Suffolk, Va.: Joint Staff J7, July 2013.

Figure A.1
Disconnect Between ISR MoPs and MoEs

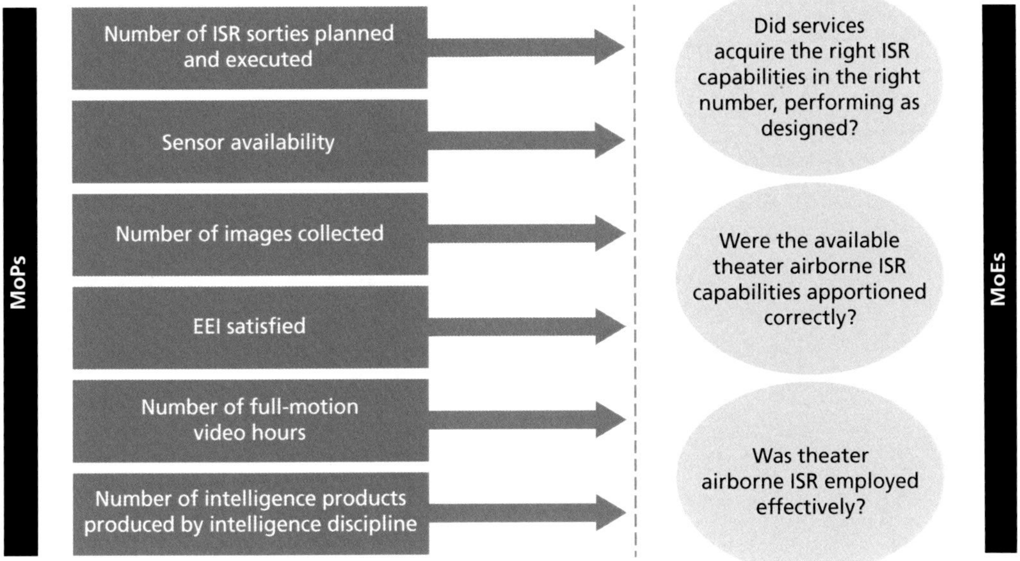

SOURCE: Adapted from Haugh and Leonard, 2017.

MoEs exists. External stakeholders may have (or acquire) additional information that can provide additional inputs, adding additional value to MoE statements. More granular system-level assessments might only require the support of three to five personnel, since those platforms are tied to more specific tasks.

Data Latency

Data latency, or the time delay between data collection, processing, and delivery to the end user, is critical within in the private sector. For example, Amazon, Target, and Wal-Mart sales rely on collecting accurate customer data and then operationalizing that data for marketing and delivering products to their customers. Technologies that assist in sales, such as real-time advertising (or real-time "bidding") require relevant and timely search history data to target user preference.[31] Open-source literature on ISR effectiveness in military contexts is limited to tactical-level capabilities and assets, although there is increasingly exploratory research around the future implications of integrating artificial intelligence and machine learning with ISR assets and data processing systems.[32]

---

[31] "Media Buying 101: What is Real-Time Bidding?" Mobidea Academy blog, May 16, 2018.

[32] For examples, see Richard M. Buchter, "2020: Faster Than Real-Time Tactical Intelligence, Surveillance, and Reconnaissance (ISR) from the Dismount, and Faster Than Real-Time Strategic ISR to the Dismount," SPIE Digital Library, May 4, 2018; Master Sgt. Heidi West, USAF, "Artificial Intelligence Proves Beneficial for ISR Data Interpretation," press release, April 19, 2018.

Data latency is critical when viewed through the lens of ISR; the loss of time or fidelity in data transfer could cost lives. Minimizing data latency through the integration of ISR requirements (within both the proposed MoP and MoE frameworks) for ISR systems is important in cases where commanders require near real-time information to make critical decisions during F3EAD missions.

USAF Annex 2-0, *Global Integrated Intelligence, Surveillance & Reconnaissance Operations*, reiterates the importance of timeliness and accuracy of requirements in achieving tactical and operation objectives: "one of the most demanding tasks for global integrated ISR personnel during emerging crises is the need to balance requirements for accuracy with those for timeliness."[33]

Public statements from service-level leadership indicate that low latency standards for ISR missions are critical to their operational success. The public release of "Air Force 2023: Delivering Decision Advantage" provides a strategic vision for the USAF ISR enterprise and suggests the compression of the observe, orient, decide, act loop or decision cycle at all levels of operations, which is identified as a primary objective in the service's effort to advance its ISR mission over the next five years.[34] At a MilSatCom USA seminar held in June 2017, Col. Curtis Carlin, U.S. Marine Corps, from CENTCOM J6, mentioned that having low latency "as an attribute, is becoming almost as important as capacity for military customers . . . processed information must be transmitted back to forces quickly enough for them to act before the situation on the ground changes, putting a premium on low latency."[35] Distinct from "timeliness," a commonly cited MoP for ISR systems, low-latency requirements call for efficacy that could be measured in time or characterized as a measure of "decision advantage."[36]

Timeliness of data is important throughout the public sector. For example, the National Oceanic and Atmospheric Administration (NOAA) regularly conducts reconnaissance missions to monitor weather patterns and events. In its "National Hurricane Operations Plan,"[37] NOAA describes requirements and criteria for what is considered a successful mission. Criteria for mission success are specific to the type of task being conducted but include timeliness as an MoP and whether or not the mission requirements had been satisfied as an MoE.

---

[33] The annex lists other important features of accurate data, including relevance (data tailored to the requestor's requirements) and that integrated ISR-derived information must be "readily accessible" and "usable." USAF Annex 2-0, 2015.

[34] Robert P. Otto, USAF, "Air Force ISR 2023: Delivering Decision Advantage," USAF, 2013.

[35] Buchter, 2018.

[36] One Naval Postgraduate School dissertation confirms that "situational awareness of the battlefield needs to be achieved through effective and proper integrated ISR network, where its effectiveness is determined by its utility to decision superiority" (Sze Shiang Soh, *Determining Intelligence, Surveillance and Reconnaissance (ISR) System Effectiveness, and Integration as Part of Force Protection and System Survivability*, dissertation, Monterey, Calif.: Naval Postgraduate School, 2013).

[37] NOAA, "National Hurricane Operations Plan," May 2018.

- **Finding:** Data latency effect can inhibit typical decisionmaking and acquisition feedback loops. Assessments that aim to inform either equipment acquisition or resource planning will require a different set of forward-looking metrics, perhaps reliant on alternative streams of data, than the metrics used to assess current operations.

Generalizing and Consolidating Disparate Metrics to Increase Assessment Utility

Developing ten simple metrics that are measurable, relevant, and clearly communicable across stakeholders is more useful than developing 100 platform-specific ones; however, defining metrics in this way can be challenging.[38] Grouping similar metrics (as discussed throughout this report) to optimize assessment mechanisms is well grounded in the literature reviewed.

Individual metrics are often combined to supply more useful information, creating metrics that evaluate programs or processes at a higher level (e.g., strategic or operational) and are not typically used in isolation. Context for the interpretation and evaluation of generalized ISR metrics is another important consideration when rolling up disparate measures.[39] For example, within the HRI literature, metrics should be interpreted only in the right context, "with respect to expectations of performance for a particular task, activity plan," and mode of operation.[40] This means that if the expected reconnaissance robot performance is different in autonomous versus remotely operated modes, it may not be reasonable to judge the summed performance by a single standard. Defining performance standards according to formal or system requirements, validation test results, or comparison to similar or alternative systems has also shown to be beneficial in producing quality metrics.[41]

---

[38] A previous RAND technical report stated that "ISR objectives must be written in a way that can be measured; otherwise, their satisfaction will always be in doubt." The authors stated that if the MoEs are specified for each measurable ISR task, and if those tasks are in turn integrated with objectives, it will be more straightforward to determine whether a task has been accomplished and its objectives are being supported. The RAND authors also suggested that the standardization of training would improve ISR task writing when needed (Lingel et al., 2008, p. 14).

[39] The National Aeronautics and Space Administration and its associated academic institutions conduct research into robotic reconnaissance for its planetary exploration programs. These reconnaissance operations have considerable parallels to military ISR activities, particularly in regard to human-robot interaction (HRI) and the challenges therein. A recurring theme is the importance of context, both for how the data of the metrics should be judged and how and by whom they should be used.

[40] Debra Schreckenghost, Terrence Fong, Tod Milam, Estrellina Pacis, and Hans Utz, "Real-Time Assessment of Robot Performance During Remote Exploration Operations," 2009 Institute of Electrical and Electronics Engineers Aerospace Conference, Big Sky, Mont., March 7–14, 2009.

[41] This could be accomplished by using algorithms to mathematically combine component metrics (often weighted by user-determined importance) to generate a new metric, such as *composite performance scores*. Alternatively, metrics could be more intuitively combined through simple ratios. For example, *productive time* could be combined with *overhead time* (defined as time spent "waiting for a reconnaissance plan or handling problems") to create a *Work Efficiency Index* (productive time:overhead time). See Debra Schreckenghost, Terrence Fong, Hans

The HRI literature we reviewed identified more than 40 frequently referenced metrics for evaluating HRI.[42] While a standard framework for the categorization of HRI metrics has not yet been developed, the literature provided options for organizing metrics by task,[43] by the object being measured,[44] and by the purpose of the end user. Figure A.2 divides HRI metrics by the object being directly measured (human, system, robot) and then further subdivides the systems (how well humans and robots perform together as a team) into productivity, efficiency, reliability, safety, and coactivity.

The taxonomy of metrics listed in Figure A.2 is purposefully general. In practice, such metrics as *trust*, *effort*, and *effectiveness* are all defined and measured according to the environment in which they are used. Ideally, metrics in the HRI field should be (1) simple and interpretable by different users, (2) focused on the functional performance of a task, (3) supported by available technology, and (4) requirements-based.[45]

HRI frameworks could be used to assist in the identification of specific classes of ISR metrics, helping to facilitate comparison of results across platforms and operations.[46] Within HRI, similar metrics are often implemented in different ways and may not always fit within a single category.[47] In addition, metric classes are often interconnected. Within the HRI field, the usefulness of a robot could also be measured by the effect on its human operator:

> Robots are not conscious, they have no projects of their own other than those assigned to them. Clancey points this out to illustrate that it's too soon to talk about human-robot cooperation or collaboration: instead, robots serve as assistants to people working toward a project goal. Therefore, the measure of a robot's

---

Utz, and Tod Milam, "Measuring Robot Performance in Real-Time for NASA Robotic Reconnaissance Operations," *Proceedings of the 9th Workshop on Performance Metrics for Intelligent Systems*, Association for Computing Machinery, September 2009.

42 Robin Murphy, and Debra Schreckenghost, "Survey of Metrics for Human-Robot Interaction," *Proceedings of the 8th ACM/IEEE International Conference on Human-Robot Interaction*, Institute of Electrical and Electronics Engineers Press, 2013.

43 Aaron Steinfeld, Terrence Fong, David Kaber, Michael Lewis Jean Scholtz, Alan Schultz, and Michael Goodrich, "Common Metrics for Human-Robot Interaction," *Proceedings of the 1st Association for Computing Machinery Special Interest Group on Computer Human Interaction/Special Interest Group on Artificial Intelligence Conference on Human-Robot Interaction*, Salt Lake City, Utah, March 2–3, 2006.

44 Schreckenghost et al. (2009) also evaluates reconnaissance robots by their *productive time*, which is defined as the amount of time the robot spends performing planned tasks. Other metrics, such as *percentage distance complete*, have different meanings or uses to those with different purposes: To the rover operator, it indicates quality of plan execution—but the science analyst would use the same metric to determine progress on data collection. See Schreckenghost et al., 2009, and Murphy and Schreckenghost, 2013.

45 Edward Tunstel, "Operational Performance Metrics for Mars Exploration Rovers," *Journal of Field Robotics*, Vol. 24, No. 8/9, 2007, pp. 651–670.

46 Steinfeld et al., 2006.

47 Murphy and Schreckenghost, 2013.

Figure A.2
Example Taxonomy of HRI Metrics

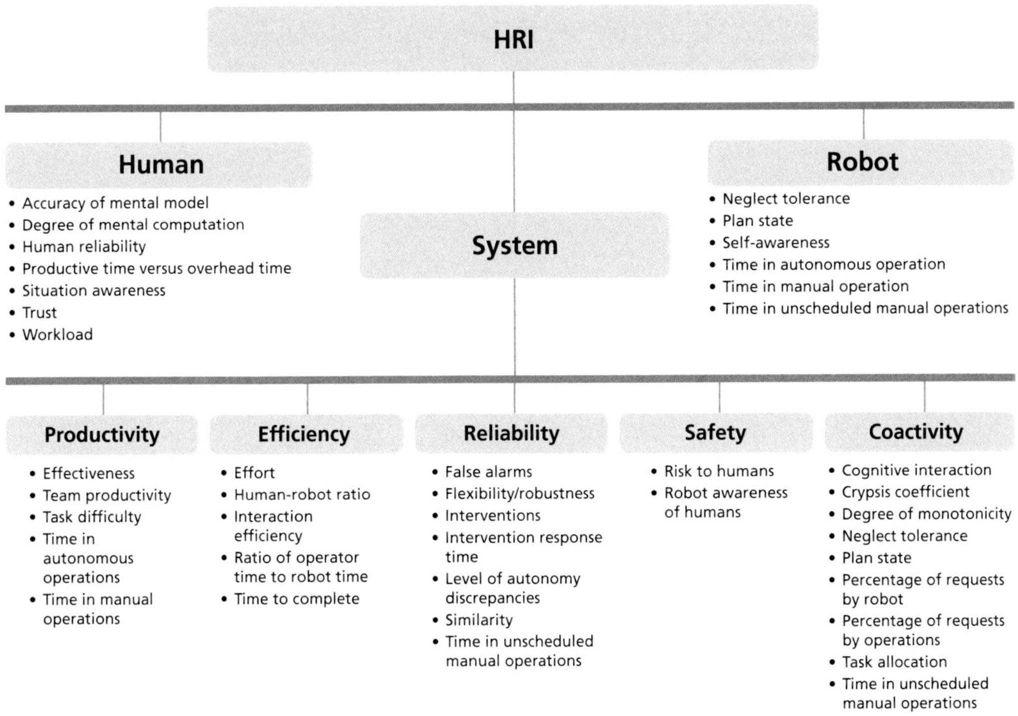

SOURCE: Murphy and Schreckenghost, 2013.

> usefulness, efficiency and functionality is based solely on whether it contributes to helping a person (or team) accomplish a goal by making that person's or team's task performance more efficient, effective, or easy in some way. This means measuring human performance (aided by robots) is the key. . . . We believe human-robot systems must be examined and measured in terms of their effect on human performance, since that is what they are designed to augment or improve.[48]

Current disparate ISR metrics that lend themselves to consolidation (meeting metric goals similar within the HRI literature) are more likely to be measurable, relevant, and understood throughout ISR plans and operations.

- **Finding:** Consolidating a vast array of metrics or data elements into more discrete, readily measurable data elements can inform a variety of assessments across organizations and operations. The unique nature of disparate measures, however,

[48] Jennifer L. Burke, Robin R. Murphy, Dawn R. Riddle, and Thomas Fincannon, "Task Performance Metrics in Human-Robot Interaction: Taking a Systems Approach," *Proceedings of the 2004 Performance Metrics for Intelligent Systems Workshop*, Gaithersburg, Md., August 24–26, 2004.

can be lost after consolidating discrete metrics or data elements. It is difficult to plan or evaluate results with measures containing multiple discrete aspects. Moreover, the use of particular former data elements to inform the newly consolidated metric will likely morph during this process. Despite such difficulties, the literature suggests that the consolidation of similar metrics will enable more effective ISR assessment processes.

Conclusions from the Literature

In our literature review, we uncovered several studies that revealed considerable commonality among the concepts identified in DoD ISR doctrine. We identified significant differences, however, in assessment mechanisms at the data analysis level. Three commonalities emerged throughout the literature. First, higher-level assessment strategies usually require inputs from multiple stakeholders (usually external to the plans division), while system-level assessments (such as those performed by the CENTCOM J2) might require the support of only three to five personnel. Second, assessments that aim to inform either equipment acquisition or resource planning will require a different set of forward-looking metrics than the metrics used to assess current operations. Third, similar data elements may inform a variety of assessments across organizations and operations, but how those data elements are used will vary. Here, the consolidation of similar metrics will assist ISR assessment processes moving forward.

When aligning program requirements to program assessments, we found that the most pertinent finding was that MoPs must focus on outputs (usually a quantitative measure, or binary), while MoEs need to measure mission outcomes (usually qualitative). Both sets of metrics must also clearly be defined in the context of their use and should be interrogated to ensure that they can be gathered as needed (a reliability measure), be accurate (a validity measure), and ensure that the metric design focuses on data and systems currently in use (a feasibility measure).

Generalizing and consolidating metrics will improve the overall assessment process. The proliferation of unique ISR platforms has increased in the post-2001 era, creating the perception that each platform must therefore require its own unique set of metrics to gauge success. As simple metrics have morphed to support more complicated (or unique) ISR platforms, however, it has become more difficult to compare the effectiveness of platforms against one another. Creating sets of MoEs and MoPs should be logically simplified enough to support general assessments across the various platforms that exist to ease the burden on those collecting and making sense of the data, and to improve the timeliness and quality of information that commanders require to make decisions.

ISR Metrics: Sample Assessment Framework and Rating System

In this appendix, we provide a sample framework of the assessment and descriptions of the metrics we use to assess the ISR mission and metrics in Appendix C.

Figure B.1
Assessment Framework

Mission (n.) Roles (n.n.) Sub-roles (n.n.n.) Activity	MoE(s)	Sub-MoE(s)	System MoP	Rule[a]	Validity	Reliability	Feasibility
[Mission description]	[Mission MoE(s)]	1. (Sub-MoE)					
		2. (Sub-MoE)					
		3. (Sub-MoE)					
1. [Role description]	1. [Role MoE]	1. (Sub-MoE)					
		2. (Sub-MoE)					
		3. (Sub-MoE)					
1.1 [Sub-role description]	1. [Sub-role MoE]	1. (Sub-MoE)	*				
		2. (Sub-MoE)	*				
		3. (Sub-MoE)	*				
1.1.1. [Activity description]	1.[Activity MoE]	1. (Sub-MoE)	*				
		2. (Sub-MoE)	*				
		3. (Sub-MoE)	*				

NOTES: MoPs are calculated at various levels to assess performance of different levels of the mission.
[a] Rules to be determined by user.

Table B.1
Rating System Metric Descriptions

Rating	Validity (ability to capture the output or outcome being assessed)	Reliability (how consistently the measurement can be made)	Feasibility (how easily the measurement can be made)
High	The metric directly measures the element.	Quantitative, well-defined, and stable	Required data sets are available and well organized.
Medium	The metric is closely related to the element (may include proxies that are closely correlated with the element).	Either qualitative, well-defined, and stable OR quantitative, less well-defined, and somewhat volatile	Required data sets could be collected without much difficulty.
Low	The metric is indirectly related to the element.	Qualitative, dependent on judgment, and anecdotal	Required data sets would be challenging to collect.

SOURCE: Scott Savitz, Henry H. Willis, Aaron Davenport, Martina Melliand, William Sasser, Elizabeth Tencza, and Dulani Woods, *Enhancing U.S. Coast Guard Metrics*, Santa Monica, Calif.: RAND Corporation, RR-1173-USCG, 2015.

APPENDIX C

CENTCOM ISR Mission and Metrics Assessment

In this appendix, we provide a sample framework of the assessment and descriptions of the metrics we use to assess the ISR mission and metrics in Appendix B.

Definitions

MoE: criterion used to assess changes in system behavior, capability, or operational environment that is tied to measuring the attainment of an end state, achievement of an objective, or creation of an effect. Denotes the effectiveness of the mission to stay on task or reduce uncertainty. Changes due to dynamic retaskings via the 8-line process will not be deemed ineffective.

Mission MoE(s): mission outcome.

Sub-MoE(s): mission or system outcome; many one-to-one or -to-many relationship to Mission MoE.

MoP: criterion used to assess friendly actions that are tied to measuring task accomplishment. Denotes performance information, e.g., deviations due to weather, maintenance, operational constraints. Does not denote the effectiveness of the mission, merely the performance of the collection and PED capability.

System MoP(s): system output; not a one-to-one relationship with MoEs.

Data element(s): data required to support metric. (Not a one-to-one relationship with MoPs and MoEs.)

The MoPs in the left column of Table C.1 are aggregated to generate the Aircraft, Collection, and Tasking Scores using decision rules, which are determined by the user. These scores are then combined to assess System MoPs of the lowest-level functions (activities and sub-roles). The System MoPs, in turn, combine to assess MoEs.

Table C.1
Calculation of System MoP

MoP	Category	
What was the sortie satisfaction rate? What was the on watch rate? What was the operational readiness rate?	Aircraft Score	System MoP
What was the percentage of standing requirements that were satisfied? What was the percentage of ad hoc requirements that were satisfied? What was the percentage of time that the sensor was operational?	Collection Score	
What was the percentage of validated requirements that were not tasked? What was the percentage of tasking denied because of lack of available platform? What was the percentage of tasking denied because of lack of available sensor? What was the percentage of occurrences in which the first-choice sensor was unavailable?	Tasking Score	

Table C.2 provides our ISR assessment framework. The different performance levels of the mission are assessed by MoPs, which are calculated at various levels. Scores for individual sortie data are first aggregated for the lowest-level components (activities when present, and sub-roles if no activities are present) to provide the final scores. For example, Role 5 ISR Support to Intel, Sub-Role 3 PMESII systems (Sub-role 5.3) consists of Activities 5.3.1–5.3.6. Therefore, the sortie scores roll up to the activity level for this role/sub-role. Whereas Role 2 ISR Support to Ops, Dynamic Targeting (Sub-Role 2.2) has no activities, and thus the sortie scores roll up at the sub-role level. Other business rules are then applied to the three scores in order to generate a single semi-quantitative assessment of the System MoP. The lowest level at which the system level is assessed is identified with an asterisk (*). System MoPs can be assessed for higher levels (sub-roles, roles) by the application of more business rules, or by simple averaging of the components. Sub-MoEs were included only when the Mission MoE was not sufficiently comprehensive.

Table C.2
ISR Assessment Framework

Mission (n.) Roles (n.n.) Sub-Roles (n.n.n.) Activity	Mission MoE(s)	Sub-MoE(s)
Mission: Intelligence, Surveillance, and Reconnaissance (ISR) is an activity that synchronizes and integrates the planning and operation of sensors, assets, and processing exploitation, and dissemination systems in direct support of current and future operations. This is an integrated intelligence and operations function.	1. To what degree did ISR synchronize and integrate the planning and operation of sensors, assets, and processing exploitation, and dissemination systems in direct support of current and future operations?	1. To what degree did ISR contribute to the mitigation of hostile actions against U.S. coalition, resources, facilities and critical infrastructure (Role 1)? 2. To what degree did ISR maintain overwatch and force protection of maneuvering friendly forces throughout the operation (Role 2)? 3. To what degree did ISR support the development and refinement of HVI target nominations (Role 3)? 4. To what degree did ISR support the development and refinement of KNK target nominations for the JTCB, operational strike planning, and BDA (Role 4)? 5. To what degree did ISR support intelligence production, JIPOE, and Warning (Role 5)?
1. ISR Support to Fixed Point Security (FPS): Support to surveillance in the immediate vicinity of a fixed location with a permanent or semi-permanent presence of coalition forces, e.g., garrisons, forward operating bases (FOBs), aerial or seaports of debarkation (APODs), etc. Preventative measures taken to mitigate hostile actions against U.S., Coalition, resources, facilities, and critical infrastructure.	1. To what degree did ISR contribute to the mitigation of hostile actions against U.S., coalition, resources, facilities, and critical infrastructure?	1. To what degree did ISR contribute to the mitigation of hostile actions by agents, saboteurs, sympathizers, terrorists, or civil disturbances against U.S., Coalition, resources, facilities and critical infrastructure (Sub-Role 1.1)? 2. To what degree did ISR contribute to the mitigation of hostile actions by small tactical units against U.S. coalition, resources, facilities and critical infrastructure (Sub-Role 1.2)? 3. To what degree did ISR contribute to the mitigation of hostile actions by large tactical force operations against U.S. coalition, resources, facilities and critical infrastructure (Sub-Role 1.3)?
1.1. Agents, saboteurs, sympathizers, terrorists, civil disturbances.	1. To what degree did ISR contribute to the mitigation of hostile actions by agents, saboteurs, sympathizers, terrorists, or civil disturbances against U.S. coalition, resources, facilities, and critical infrastructure?	1. To what degree did ISR tasking and collection contribute to the identification of threats or plots by agents, saboteurs, sympathizers, terrorists, and civil disturbances?* 2. To what degree did ISR tasking and collection contribute to the identification of agent, saboteur, sympathizer, terrorist, civil disturbance networks?* 3. To what degree did ISR tasking and collection contribute to the verification of agent, saboteur, sympathizer, terrorist, civil disturbance networks?*
1.2. Small tactical units; irregular forces may include significant stand-off weapons threats.	1. To what degree did ISR contribute to the mitigation of hostile actions by small tactical units against U.S. coalition, resources, facilities, and critical	1. To what degree did ISR tasking and collection contribute to the identification of small tactical units?* 2. To what degree did ISR identify irregular forces?* 3. To what degree did ISR identify significant stand-off weapon threats?*

Table C.2—Continued

Mission (n.) Roles (n.n.) Sub-Roles (n.n.n.) Activity	Mission MoE(s)	Sub-MoE(s)
1.3. Large tactical force operations, including airborne, heliborne, amphibious, infiltration, and major air operations. Area reconnaissance for rear area and base security are assessed to be "economy of force measures" and so are considered to be of lesser priority.	1. To what degree did ISR contribute to the mitigation of hostile actions by large tactical force operations against U.S. coalition, resources, facilities, and critical infrastructure?	1. To what degree did ISR tasking and collection contribute to the identification of large tactical airborne forces?*
		2. To what degree did ISR identify large tactical heliborne forces?*
		3. To what degree did ISR identify large tactical amphibious forces?*
		4. To what degree did ISR identify large tactical infiltration forces?*
		5. To what degree did ISR identify large tactical air operations forces?*
2. ISR Support to Operations (Ops): Support to overwatch (OW) and force protection of maneuvering friendly forces (land, sea, or air); includes dynamic targeting support to units conducting advise and assist missions.	1. To what degree did ISR maintain OW and force protection of maneuvering friendly forces (land, sea, or air) throughout the operation (to include dynamic targeting support to units conducting advise and assist missions)?	1. To what degree did ISR maintain overwatch of maneuvering friendly forces (Sub-Role 2.1)?
		2. To what degree did ISR maintain dynamic targeting in support of maneuvering friendly forces (Sub-Role 2.2)?
2.1. Overwatch (OW): Task is to detect, locate, track, and identify entities in an assigned area and gain additional environmental data.	1. To what degree did ISR maintain OW of maneuvering friendly forces?	1. To what degree did ISR maintain OW of maneuvering friendly forces related to HVU (Activity 2.1.1)?
		2. To what degree did ISR maintain OW of maneuvering friendly forces related to NHVU (Activity 2.1.2)?
		3. To what degree did ISR maintain OW of maneuvering friendly forces related to USFOR (Activity 2.1.3)?
		4. To what degree did ISR maintain OW of maneuvering friendly forces related to AMCIT (Activity 2.1.4)?
2.1.1. High Value Unit (HVU)	1. To what degree did ISR maintain OW of maneuvering friendly forces related to HVU?	1. To what degree did ISR maintain OW of maneuvering HVU land forces?*
		2. To what degree did ISR maintain OW of maneuvering HVU sea forces?*
		3. To what degree did ISR maintain OW of maneuvering HVU air forces?*
2.1.2. Non-High Value Unit (NHVU)	1. To what degree did ISR maintain OW of maneuvering friendly forces related to HVU?	1. To what degree did ISR maintain OW of maneuvering HVU land forces?*
		2. To what degree did ISR maintain OW of maneuvering HVU sea forces?*
		3. To what degree did ISR maintain OW of maneuvering HVU air forces?*
2.1.3. U.S. Forces (USFOR)	1. To what degree did ISR maintain OW of maneuvering USFOR friendly forces?	1. To what degree did ISR maintain OW of maneuvering USFOR land forces?*
		2. To what degree did ISR maintain OW of maneuvering USFOR sea forces?*
		3. To what degree did ISR maintain OW of maneuvering USFOR air forces?*

Table C.2—Continued

Mission (n.) Roles (n.n.) Sub-Roles (n.n.n.) Activity	Mission MoE(s)	Sub-MoE(s)
2.1.4. American Citizen (AMCIT)	1. To what degree did ISR maintain OW of maneuvering AMCITs?*	
2.2. Dynamic Targeting (DT)	1. To what degree did ISR maintain dynamic targeting in support of maneuvering friendly forces, to include those units conducting advise and assist missions?*	
3. ISR Support to High-Value Individual (HVI): Support to the development and refinement of HVI target nominations. Components: Find, Fix, Track, Assess.	1. To what degree did ISR support the development and refinement of HVI target nominations?	1. To what degree did ISR support the Find phase of HVI target nomination development and refinement (Sub-Role 2. To what degree did ISR support the Fix phase of HVI target nomination development and refinement (Sub-Role 3.2)? 3. To what degree did ISR support the Track phase of HVI target nomination development and refinement (Sub-Role 3.3)? 6. To what degree did ISR support the Assess phase of HVI target nomination development and refinement (Sub-Role 3.4)?
3.1. Phase 1: Find	1. To what degree did ISR support the Find phase of target nominations?*	
3.2. Phase 2: Fix	1. To what degree did ISR support the Fix phase of target nominations?*	
3.3. Phase 3: Track	1. To what degree did ISR support the Track phase of target nominations?*	
3.4. Phase 4: Assess	1. To what degree did ISR support the Assess phase of target nominations?*	
4. ISR Support to Kinetic/Non-Kinetic Targeting (KNK): Support for the development and refinement of target nominations for JTCB or like body adjudication, operational strike planning, and BDA.	1. To what degree did ISR support the development and refinement of KNK target nominations for the JTCB, operational strike planning ,and BDA?	1. To what degree did ISR support TGTDEV (Sub-Role 4.1)? 2. To what degree did ISR support BDA (Sub-Role 4.2)?

Table C.2—Continued

Mission / (n.) Roles / (n.n.) Sub-Roles / (n.n.n.) Activity	Mission MoE(s)	Sub-MoE(s)
4.1. Deliberate Target Development (TGTDEV)	1. To what degree did ISR contribute to the analysis, assessment, and documentation processes used to identify and characterize potential targets that, when engaged, support achievement of the Commander's objective?	1. To what degree did ISR contribute to the TDN phase of TGTDEV (Activity 4.1.1)? 2. To what degree did ISR contribute to the POA phase of TGTDEV (Activity 4.1.2)? 3. To what degree did ISR contribute to the POL phase of TGTDEV (Activity 4.1.3)? 4. To what degree did ISR contribute to the PID phase of TGTDEV (Activity 4.1.4)?
4.1.1. Phase 1: Target Development Nomination (TDN)	1. To what degree did ISR contribute to the TDN phase of TGTDEV?*	
4.1.2. Phase 2: Pattern of Activity (POA)	1. To what degree did ISR contribute to the POA phase of TGTDEV?*	
4.1.3. Phase 3: Pattern of Life (POL)	1. To what degree did ISR contribute to the POL phase of TGTDEV?*	
4.1.4. Phase 4: Positive Identification (PID)	4. To what degree did ISR contribute to the PID phase of TGTDEV?*	
4.2. Battle Damage Assessment (BDA)	1. What outcomes were achieved by ISR support in the three phases of BDA: physical damage / change assessment, functional damage assessment, and target system assessment)?	1. To what degree did ISR contribute to ITA of BDA (Activity 4.2.1)? 2. To what degree did ISR contribute to STA of BDA (Activity 4.2.2)? 3. To what degree did ISR contribute to TSA of BDA (Activity 4.2.3)?
4.2.1. Initial Target Assessment (ITA)	1. What amount of ITA information for BDA was supplied by ISR?	1. To what degree did ISR support physical damage/change assessment?* 2. To what degree did ISR support functional damage assessment?* 3. To what degree did ISR support target system assessment?*
4.2.2. Supplemental Target Assessment (STA)	1. What amount of STA information for BDA was supplied by ISR?	1. To what degree did ISR support physical damage/change assessment?* 2. To what degree did ISR support functional damage assessment?* 3. To what degree did ISR support target system assessment?*
4.2.3. Target System Assessment (TSA)	1. What amount of TSA information for BDA was supplied by ISR?	1. To what degree did ISR support physical damage/change assessment?* 2. To what degree did ISR support functional damage assessment?* 3. To what degree did ISR support target system assessment?*

Table C.2—Continued

Mission / (n.) Roles / (n.n.) Sub-Roles / (n.n.n.) Activity	Mission MoE(s)	Sub-MoE(s)
5. ISR Support to Intelligence (INTEL): Support to intelligence production, Joint Intelligence Preparation of the Operational Environment (JIPOE), and Warning in the three domains (physical (air, land, maritime, and space), information environment (including cyberspace), and political, military, economic, social, information, and infrastructure (PMESII) systems and subsystems).	1. To what degree did ISR provide support to the relevant aspects of the environment, including the adversary and other actors, for intelligence production, JIPOE, and Warning in the three domains (physical, information environment, and PMESII systems and subsystems)?	1. To what degree did ISR provide support to intelligence production, JIPOE, and Warning in the physical domains (Sub-Role 5.1)? 2. To what degree did ISR provide support to intelligence production, JIPOE, and Warning in the information environment (Sub-Role 5.2)? 3. To what degree did ISR provide support to intelligence production, JIPOE, and Warning for PMESII systems and subsystems (Sub-Role 5.3)?
5.1. Physical domains	1. To what degree did ISR support the physical domains (air, land, maritime, and space) to intelligence production, JIPOE, and Warning?	1. To what degree did ISR provide support to intelligence production, JIPOE, and Warning in the air domain (Activity 5.1.1)? 2. To what degree did ISR provide support to intelligence production, JIPOE, and Warning in the land domain (Activity 5.1.2)? 3. To what degree did ISR provide support to intelligence production, JIPOE, and Warning in the maritime domain (Activity 5.1.3)? 4. To what degree did ISR provide support to intelligence production, JIPOE, and Warning in the space domain (Activity 5.1.4)?
5.1.1. Air domain	1. To what degree did ISR provide support to intelligence production, JIPOE, and Warning in the air domain?*	
5.1.2. Land domain	1. To what degree did ISR provide support to intelligence production, JIPOE, and Warning in the land domain?*	
5.1.3. Maritime domain	1. To what degree did ISR provide support to intelligence production, JIPOE, and Warning in the maritime domain?*	
5.1.4. Space domain	1. To what degree did ISR provide support to intelligence production, JIPOE, and Warning in the space domain?*	
5.2. Information environment (including cyberspace)	1. To what degree did ISR provide support to intelligence production, JIPOE, and Warning in the information environment (including cyberspace)?*	

Table C.2—Continued

Mission / (n.) Roles / (n.n.) Sub-Roles / (n.n.n.) Activity	Mission MoE(s)	Sub-MoE(s)
5.3. PMESII systems and subsystems	1. To what degree did ISR provide support to intelligence production, JIPOE, and Warning for PMESII systems and subsystems?	1. To what degree did ISR provide support to intelligence production, JIPOE, and Warning for political systems and subsystems (Activity 5.3.1)?
		2. To what degree did ISR provide support to intelligence production, JIPOE, and Warning for military systems and subsystems (Activity 5.3.2)?
		3. To what degree did ISR provide support to intelligence production, JIPOE, and Warning for economic systems and subsystems (Activity 5.3.3)?
		4. To what degree did ISR provide support to intelligence production, JIPOE, and Warning for social systems and subsystems (Activity 5.3.4)?
		5. To what degree did ISR provide support to intelligence production, JIPOE, and Warning for information systems and subsystems (Activity 5.3.5)?
		6. To what degree did ISR provide support to intelligence production, JIPOE, and Warning for infrastructure systems and subsystems (Activity 5.3.6)?
5.3.1. Political systems and subsystems	1. To what degree did ISR provide support to intelligence production, JIPOE, and Warning for political systems and	
5.3.2. Military systems and subsystems	1. To what degree did ISR provide support to intelligence production, JIPOE, and Warning for military systems and	
5.3.3. Economic systems and subsystems	1. To what degree did ISR provide support to intelligence production, JIPOE, and Warning for economic systems and	
5.3.4. Social systems and subsystems	1. To what degree did ISR provide support to intelligence production, JIPOE, and Warning for social systems and	
5.3.5. Information systems and subsystems	1. To what degree did ISR provide support to intelligence production, JIPOE, and Warning for information systems and subsystems?*	
5.3.6. Infrastructure systems and subsystems	1. To what degree did ISR provide support to intelligence production, JIPOE, and Warning for infrastructure systems and subsystems?*	

APPENDIX D

Metrics Visualization

In this appendix, we provide screenshots of the dashboards not included in the main report for the interactive, Tableau-based metrics visualization tool.

Figure D.1
Map Dashboard

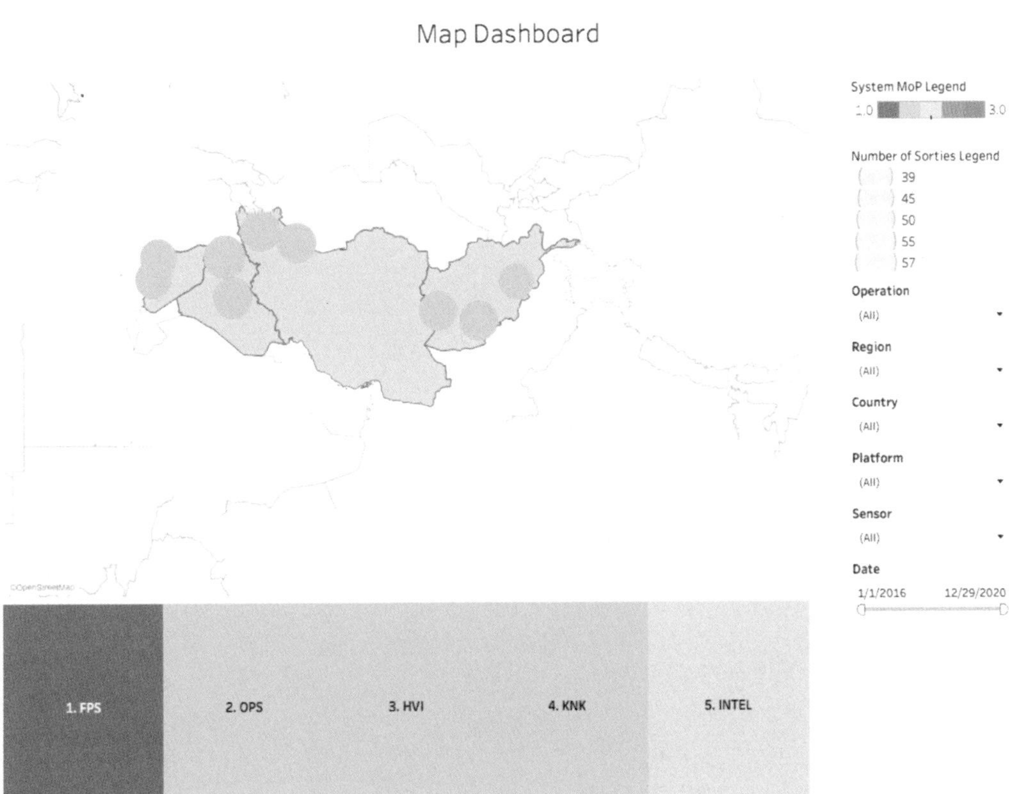

Figure D.2
Platform Comparison Dashboard

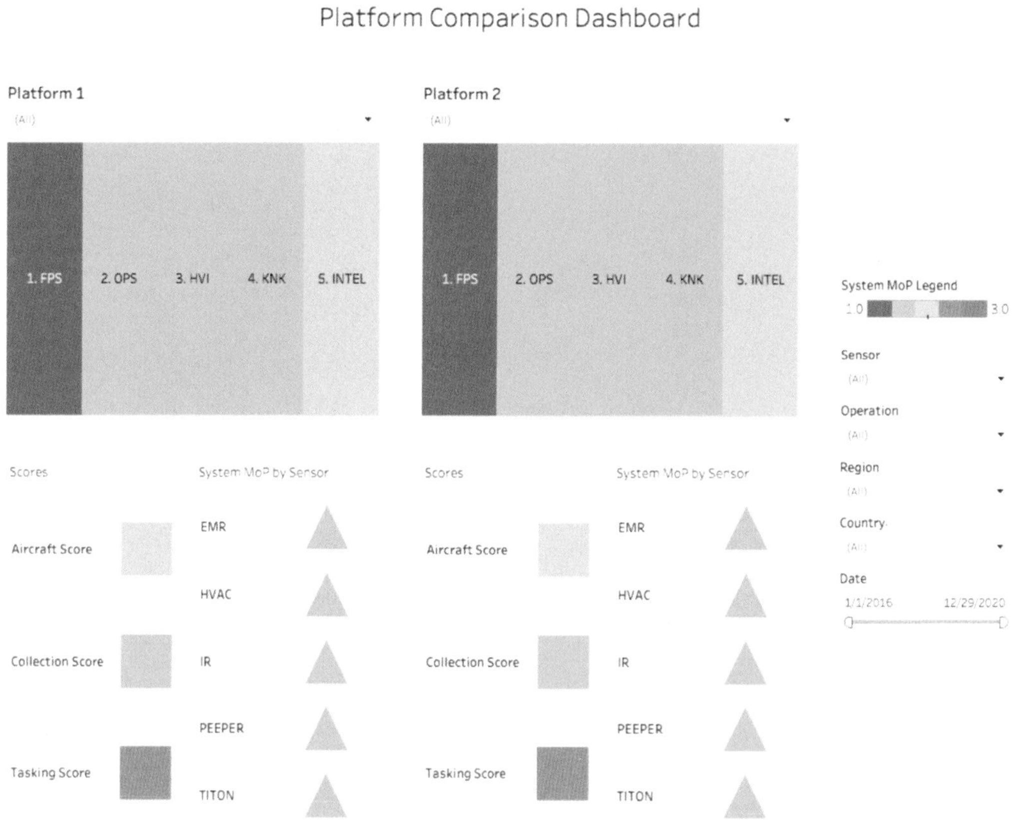

Figure D.3
Sensor Comparison Dashboard

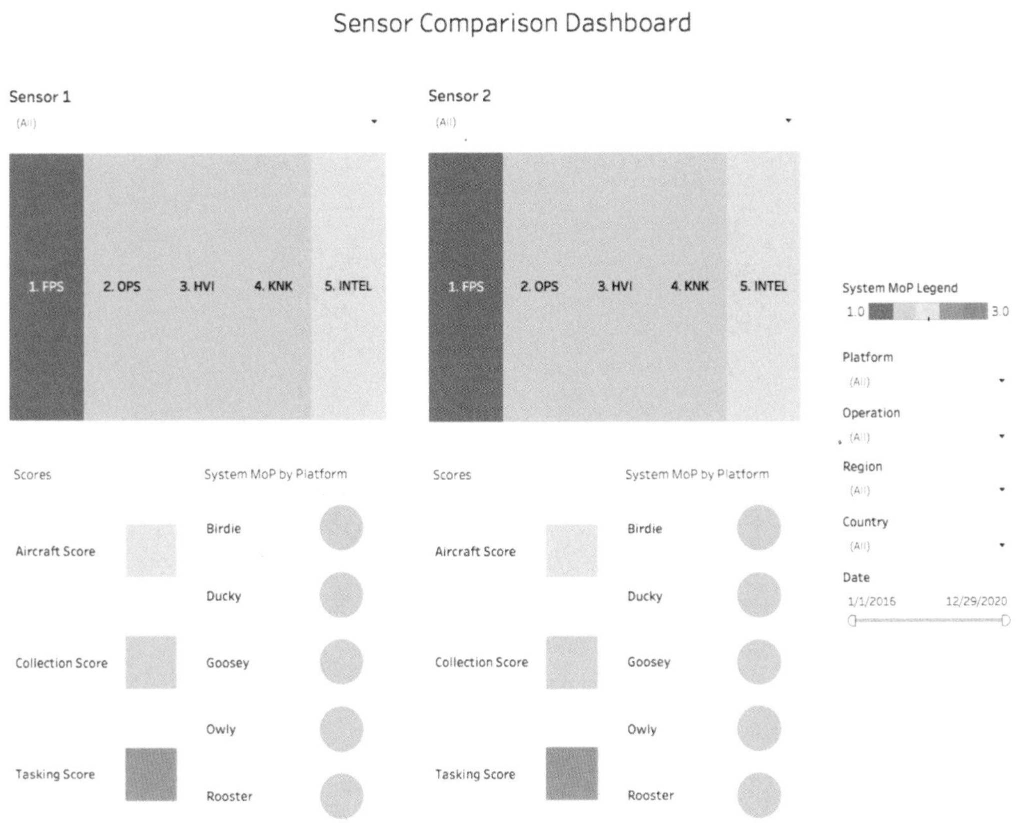

Figure D.4
Missing Data Dashboard

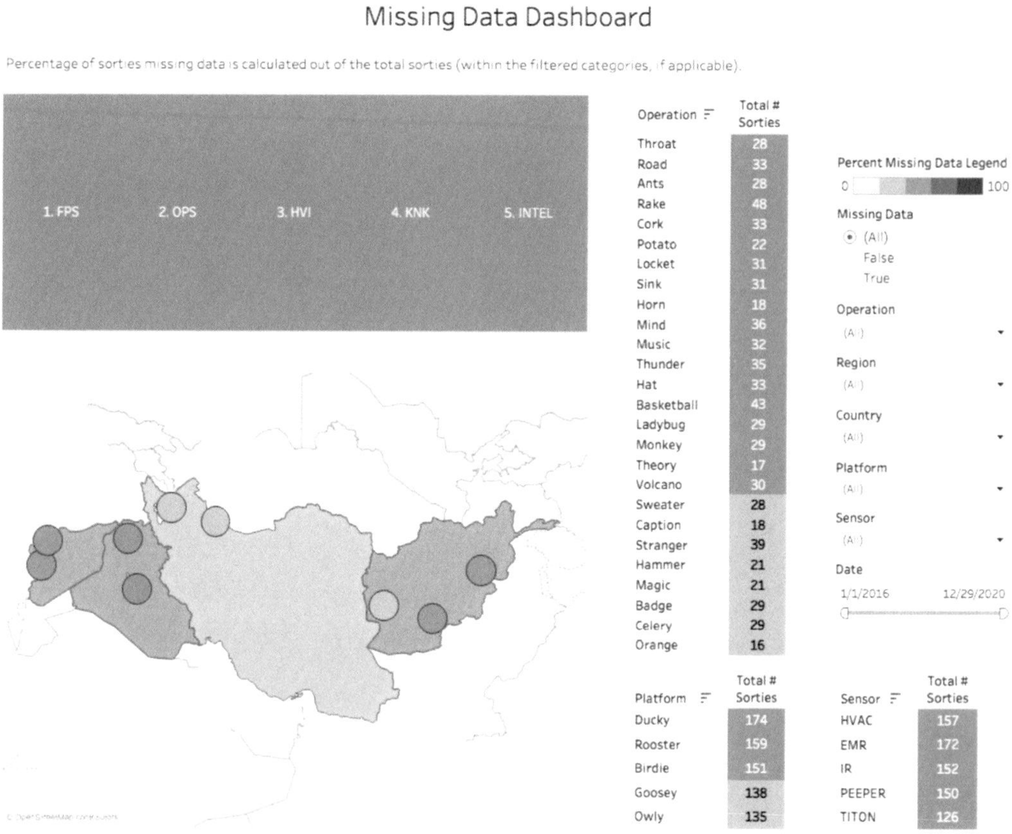

References

Army Doctrine Publication 3-37, *Protection*, July 2019, supersedes Army Doctrine Reference Publication 3-37, August 2012. As of October 23, 2019:
https://fas.org/irp/doddir/army/adp3_37.pdf

Army Technical Publication 3-60, *Targeting*, May 2015. As of October 23, 2019:
https://fas.org/irp/doddir/army/atp3-60.pdf

Buchter, Richard M., "2020: Faster Than Real-Time Tactical Intelligence, Surveillance, and Reconnaissance (ISR) from the Dismount, and Faster Than Real-Time Strategic ISR to the Dismount," SPIE Digital Library, May 4, 2018.

Burke, Jennifer L., Robin R. Murphy, Dawn R. Riddle, and Thomas Fincannon, "Task Performance Metrics in Human-Robot Interaction: Taking a Systems Approach," *Proceedings of the 2004 Performance Metrics for Intelligent Systems Workshop*, Gaithersburg, Md., August 24–26, 2004. As of March 30, 2020:
https://pdfs.semanticscholar.org/fa60/c4e8145d9ae61988e3a38ca22f7b9c71ea09.pdf

Chairman of the Joint Chiefs of Staff Instruction 3162.02, *Methodology for Combat Assessment*, Washington, D.C., March 8, 2019. As of March 30, 2020:
https://www.jcs.mil/Portals/36/Documents/Library/Instructions/CJCSI%203162.02.pdf?ver=2019-03-18-121700-237

Chairman of the Joint Chiefs of Staff Instruction 3340.02B, *Joint Enterprise Integration of Warfighter Intelligence*, Washington, D.C., October 24, 2013. As of March 30, 2020:
https://www.jcs.mil/Portals/36/Documents/Library/Instructions/3340_02.pdf?ver=2016-02-05-175030-873

Chairman of the Joint Chiefs of Staff Manual 33154.01A, *Intelligence Planning*, Washington, D.C., September 17, 2012. As of March 30, 2020:
https://info.publicintelligence.net/CJCS-IntelPlanning.pdf

CJCSI—*See* Chairman of the Joint Chiefs of Staff Instruction.

CJCSM—*See* Chairman of the Joint Chiefs of Staff Manual.

Commander's Handbook for Assessment Planning and Execution, version 1.0, Suffolk, Va.: Joint Staff J-7, Joint and Coalition Warfighting, September 9, 2011a.

Commander's Handbook for Persistent Surveillance, version 1.0, Suffolk, Va.: Joint Warfighting Center, U.S. Joint Forces Command, Joint Doctrine Support Division, June 20, 2011b.

Cordova, Amado, Kirsten M. Keller, Lance Menthe, and Carl Rhodes, *Virtual Collaboration for a Distributed Enterprise*, Santa Monica, Calif.: RAND Corporation, RR-153-AF, 2013. As of October 15, 2013:
http://www.rand.org/pubs/research_reports/RR153.html

Cordova, Amado, Lindsay D. Millard, Lance Menthe, Robert A. Guffey, and Carl Rhodes, *Motion Imagery Processing and Exploitation (MIPE)*, Santa Monica, Calif.: RAND Corporation, RR-154-AF, 2013. As of March 30, 2020:
https://www.rand.org/pubs/research_reports/RR154.html

Deployable Training Division, *Intelligence Operations*, 1st ed., Insights and Best Practices Focus Paper, Suffolk, Va.: Joint Staff J-7, July 2013.

GAO—*See* U.S. Government Accountability Office.

Haugh, Brig Gen Timothy D., and Lt Col Douglas W. Leonard, "Improving Outcomes: Intelligence, Surveillance, and Reconnaissance Assessment," *Air & Space Power Journal*, Winter 2017.

Lingel, Sherrill, Carl Rhodes, Amado Cordova, Jeff Hagen, Joel Kvitky, and Lance Menthe, *Methodology for Improving the Planning, Execution, and Assessment of Intelligence, Surveillance, and Reconnaissance Operations*, Santa Monica, Calif.: RAND Corporation, TR-459-AF, 2008. As of April 17, 2020:
https://www.rand.org/pubs/technical_reports/TR459.html

"Media Buying 101: What is Real-Time Bidding?" Mobidea Academy blog, May 16, 2018. As of March 30, 2020:
https://www.mobidea.com/academy/real-time-bidding/

Menthe, Lance, Amado Cordova, Carl Rhodes, Rachel Costello, and Jeffrey Sullivan, *The Future of Air Force Motion Imagery Exploitation: Lessons from the Commercial World*, Santa Monica, Calif.: RAND Corporation, TR-1133-AF, 2012. As of October 15, 2013:
http://www.rand.org/pubs/technical_reports/TR1133.html

Murphy, Robin, and Debra Schreckenghost, "Survey of Metrics for Human-Robot Interaction," *Proceedings of the 8th ACM/IEEE International Conference on Human-Robot Interaction*, Tokyo, March 3–6, 2013. As of March 30, 2020:
https://ieeexplore.ieee.org/stamp/stamp.jsp?tp=&arnumber=6483569

National Oceanic and Atmospheric Administration, "National Hurricane Operations Plan," May 2018. As of March 30, 2020:
https://www.hsdl.org/?view&did=818361

NOAA—*See* National Oceanic and Atmospheric Administration.

Office of the Chairman of the Joint Chiefs of Staff, *DOD Dictionary of Military and Associated Terms*, Washington, D.C.: The Joint Staff, last updated January 2020. As of March 30, 2020:
https://www.jcs.mil/Portals/36/Documents/Doctrine/pubs/dictionary.pdf

Otto, Robert P., U.S. Air Force, "Air Force ISR 2023: Delivering Decision Advantage," U.S. Air Force, 2013.

Savitz, Scott, Henry H. Willis, Aaron Davenport, Martina Melliand, William Sasser, Elizabeth Tencza, and Dulani Woods, *Enhancing U.S. Coast Guard Metrics*, Santa Monica, Calif.: RAND Corporation, RR-1173-USCG, 2015. As of March 30, 2020:
https://www.rand.org/pubs/research_reports/RR1173.html

Schreckenghost, Debra, Terrence Fong, Tod Milam, Estrellina Pacis, and Hans Utz, "Real-Time Assessment of Robot Performance During Remote Exploration Operations," 2009 Institute of Electrical and Electronics Engineers Aerospace Conference, Big Sky, Mont., March 7–14, 2009. As of March 30, 2020:
http://www.dtic.mil/dtic/tr/fulltext/u2/a518168.pdf

Schreckenghost, Debra, Terrence Fong, Hans Utz, and Tod Milam, "Measuring Robot Performance in Real-Time for NASA Robotic Reconnaissance Operations," *Proceedings of the 9th Workshop on Performance Metrics for Intelligent Systems*, Association for Computing Machinery, September 2009.

Soh, Sze Shiang, *Determining Intelligence, Surveillance and Reconnaissance (ISR) System Effectiveness, and Integration as Part of Force Protection and System Survivability*, dissertation, Monterey, Calif.: Naval Postgraduate School, 2013.

Steinfeld, Aaron, Terrence Fong, David Kaber, Michael Lewis Jean Scholtz, Alan Schultz, and Michael Goodrich, "Common Metrics for Human-Robot Interaction," *Proceedings of the 1st Association for Computing Machinery Special Interest Group on Computer Human Interaction/Special Interest Group on Artificial Intelligence Conference on Human-Robot Interaction*, Salt Lake City, Utah, March 2–3, 2006. As of March 30, 2020:
https://www.ri.cmu.edu/pub_files/pub4/steinfeld_aaron_m_2006_1/steinfeld_aaron_m_2006_1.pdf

Tunstel, Edward, "Operational Performance Metrics for Mars Exploration Rovers," *Journal of Field Robotics*, Vol. 24, No. 8/9, 2007, pp. 651–670. As of March 30, 2020:
https://onlinelibrary.wiley.com/doi/abs/10.1002/rob.20205

USAF—*See* U.S. Air Force.

U.S. Air Force Annex 2-0, *Global Integrated Intelligence, Surveillance & Reconnaissance Operations*, Montgomery, Ala.: Curtis E. Lemay Center, January 29, 2015. As of March 15, 2019:
https://www.doctrine.af.mil/Portals/61/documents/Annex_2-0/2-0-D01-ISR-Introduction.pdf

U.S. Central Command, "Area of Responsibility," webpage, undated. As of March 31, 2020:
https://www.centcom.mil/AREA-OF-RESPONSIBILITY/

U.S. Joint Chiefs of Staff, *Joint Targeting*, Joint Publication 3-60, January 31, 2013. As of April 17, 2020:
https://www.justsecurity.org/wp-content/uploads/2015/06/Joint_Chiefs-Joint_Targeting_20130131.pdf

U.S. Joint Chiefs of Staff, *Joint Intelligence Preparation of the Operational Environment*, Joint Publication 2-01.3, May 21, 2014a. As of March 30, 2020:
https://fas.org/irp/doddir/dod/jp2-01-3.pdf

U.S. Joint Chiefs of Staff, *Joint Fire Support*, Joint Publication 3-09, December 12, 2014b. As of March 30, 2020:
https://www.jcs.mil/Portals/36/Documents/Doctrine/pubs/jp3_09.pdf

U.S. Joint Chiefs of Staff, *Joint Planning*, Joint Publication 5-0, June 16, 2017a. As of March 30, 2020:
https://www.jcs.mil/Portals/36/Documents/Doctrine/pubs/jp5_0_20171606.pdf

U.S. Joint Chiefs of Staff, *Geospatial Intelligence in Joint Operations*, Joint Publication 2-03, July 5, 2017b. As of March 30, 2020:
https://fas.org/irp/doddir/dod/jp2_03.pdf

U.S. Joint Chiefs of Staff, *Joint and National Intelligence Support to Military Operations*, Joint Publication 2-01, July 5, 2017c. As of March 15, 2019:
https://www.jcs.mil/Portals/36/Documents/Doctrine/pubs/jp2_01_20170705v2.pdf

U.S. Government Accountability Office, *Intelligence, Surveillance, and Reconnaissance: Actions Are Needed to Increase Integration and Efficiencies of DOD's ISR Enterprise*, GAO-11-465, Washington, D.C., June 2011. As of March 30, 2020:
https://www.gao.gov/new.items/d11465.pdf

U.S. Government Accountability Office, *Intelligence, Surveillance, and Reconnaissance: DOD Can Better Assess and Integrate ISR Capabilities and Oversee Development of Future ISR Requirements*, GAO-08-374, Washington, D.C., March 2008. As of March 30, 2020: https://www.gao.gov/assets/280/273939.pdf

Votel, Joseph, "National Security Challenges and U.S. Military Activities in the Greater Middle East and Africa," testimony before the U.S. House of Representatives Armed Services Committee, Washington, D.C., March 11, 2013.

West, Heidi, U.S. Air Force, "Artificial Intelligence Proves Beneficial for ISR Data Interpretation," press release, April 19, 2018. As of March 30, 2020: https://www.af.mil/News/Article-Display/Article/1498694/artificial-intelligence-proves-beneficial-for-isr-data-interpretation/

Wicker, Amanda, Sasha Romanosky, Cortney Weinbaum, Bradley Knopp, and David Luckey, *Measuring Intelligence, Surveillance, and Reconnaissance Effectiveness at the United States Central Command: Data Visualization Tool Documentation*, Santa Monica, Calif.: RAND Corporation, TL-358-OSD, 2020.